Advanced Gear Manufacturing and Finishing

Advanced Gear Manufacturing and Finishing

Classical and Modern Processes

Kapil Gupta
Sr. Lecturer, Department of Mechanical and Industrial Engineering Technology, University of Johannesburg, Johannesburg, RSA

Neelesh Kumar Jain
Professor, Discipline of Mechanical Engineering, Indian Institute of Technology Indore, Indore, India

Rudolph Laubscher
Associate Professor, Department of Mechanical Engineering Science, University of Johannesburg, Johannesburg, RSA

Academic Press is an imprint of Elsevier
125 London Wall, London EC2Y 5AS, United Kingdom
525 B Street, Suite 1800, San Diego, CA 92101-4495, United States
50 Hampshire Street, 5th Floor, Cambridge, MA 02139, United States
The Boulevard, Langford Lane, Kidlington, Oxford OX5 1GB, United Kingdom

Notices
Knowledge and best practice in this field are constantly changing. As new research and experience broaden our understanding, changes in research methods, professional practices, or medical treatment may become necessary.

Practitioners and researchers must always rely on their own experience and knowledge in evaluating and using any information, methods, compounds, or experiments described herein. In using such information or methods they should be mindful of their own safety and the safety of others, including parties for whom they have a professional responsibility.

To the fullest extent of the law, neither the Publisher nor the authors, contributors, or editors, assume any liability for any injury and/or damage to persons or property as a matter of products liability, negligence or otherwise, or from any use or operation of any methods, products, instructions, or ideas contained in the material herein.

British Library Cataloguing-in-Publication Data
A catalogue record for this book is available from the British Library

Library of Congress Cataloging-in-Publication Data
A catalog record for this book is available from the Library of Congress

ISBN: 978-0-12-804460-5

Publisher: Matthew Deans
Acquisition Editor: Brian Guerin
Editorial Project Manager: Edward Payne
Production Project Manager: Anusha Sambamoorthy
Cover Designer: Victoria Pearson

Typeset by MPS Limited, Chennai, India

Contents

Preface

Gears and the associated gear manufacturing industry maintain a unique but significant position in the manufacturing sector at large. Almost every manufacturing industry utilizes gears and/or gear assemblies in one form or the other as part of their manufacturing process. The gear industry is not immune to technological advancement and requires new processes and techniques to fulfill the requirements of new and unique specialized transmissions that are being developed all the time. This development of new and novel techniques along with the advancements in the conventional processes needs to occur within the framework of quality, productivity, and ecological impact.

The aim of this book is to provide a concise collection of research and development aspects, salient features, applications, process principles and mechanism of advanced gear manufacturing processes.

It consists of seven chapters that include an introduction to gear engineering, conventional and advanced gear manufacturing and finishing processes, surface modification of gears, and eventually concludes with a chapter on gear metrology. Chapter 1, Introduction to Gear Engineering, introduces gears, their history, classification and type, important terminology, materials, and their manufacture. Chapter 2, Conventional Manufacturing of Cylindrical Gears, discusses conventional manufacturing of gears in brief. Chapter 3, Manufacturing of Conical and Noncircular Gears, provides an overview of conventional manufacturing of conical and non-circular gears. Chapter 4, Advances in Gear Manufacturing, is dedicated to advances in gear manufacturing and thus discusses laser machining, abrasive water jet machining, spark erosion machining, various additive processes including additive layer manufacturing, LIGA, and sustainable manufacturing of gears in more detail. Conventional and advanced gear finishing processes are presented in Chapter 5, Conventional and Advanced Finishing of Gears. Surface property enhancement techniques inclusive of gear coatings are the main focus of Chapter 6, Surface Property Enhancement of Gears. The book concludes with Chapter 7, Measurement of Gear Accuracy, where the gear metrology is presented and discussed.

The main audience targeted for this book is researchers, engineers, technical experts, and specialists working in the area of gear manufacturing and finishing.

The authors acknowledge Academic Press Inc. for this opportunity and for their professional support. Finally, the authors would like to thank all those who assisted during the development of this book.

Kapil Gupta
Sr. Lecturer, Department of Mechanical and
Industrial Engineering Technology,
University of Johannesburg, Johannesburg, RSA

Neelesh Kumar Jain
Professor, Discipline of Mechanical Engineering,
Indian Institute of Technology Indore, Indore, India

Rudolph Laubscher
Associate Professor, Department of Mechanical Engineering Science,
University of Johannesburg, Johannesburg, RSA

Chapter 1

Introduction to Gear Engineering

1.1 INTRODUCTION AND HISTORY OF GEARS

1.1.1 Introduction

A gear is basically a toothed wheel that works in tandem with another gear (or gears) to transmit power and/or motion to change speed and/or direction of motion. Dudley defined a gear as "a geometric shape that has teeth uniformly spaced around the circumference and is made to mesh its teeth with another gear" [1]. Slipping is a major problem during transmission of motion and power between two shafts by rope or belt drive and consequently may affect the precision and efficiency of the system adversely. This slipping phenomenon is largely avoided by means of gear drives. The compact layout, flexibility, high efficiency, and reliability are the most important features that make gears and gear drives the first choice in many applications. Gear sizes range from nanometers (nanogears) to meters (macrogears) with corresponding application areas from nanoelectromechanical systems (NEMS) to large mills and wind turbines. A wide range of materials ranging from plastics and ceramics to ultrahigh strength steels are used in gear manufacture.

Gears and subsequently the gear manufacturing industry plays an integral role in many industrial sectors as it is one of the basic mechanical components used for transmission of motion and/or power in equipment, machines, and instruments. Several conventional and advanced methods of gear manufacture are available for use in specialized applications to produce gears that are fit for purpose. Technological advancements in gear engineering over the last few decades have enabled the gear industry to produce near-net shaped and high-quality gears by short process chains and a lower environmental footprint.

1.1.2 History

The writings of Aristotle (4th century B.C.) reflect some of the earliest reference to gears and their use [2]. He specifically noted that the direction of rotation is reversed when one gear wheel drives another. Water-lifting devices, in the form of 'Persian wheels', were used in the 3rd century B.C. Animals such

Advanced Gear Manufacturing and Finishing. DOI: http://dx.doi.org/10.1016/B978-0-12-804460-5.00001-8

FIGURE 1.1 'Persian wheel': A water-lifting gearing mechanism used during 3rd century. *Source: Reproduced with permission from P.L. Fraenkel, Water lifting devices, FAO irrigation and Drainage Paper 43, Corporate Document Repository, Food and Agriculture Organization of the United Nations, Rome, 1986 [4].*

as camels, bullocks, and buffaloes were used to drive these devices that were typically associated with open wells. In this arrangement, an animal driven horizontal toothed wheel was meshed into a vertical toothed wheel that was then used to lift water containers that were attached to another geared mechanism (Fig. 1.1). Later on, this method was successfully adopted for use in water-driven grain mills and other devices. During the 3rd century, Archimedes also developed a device (Antikythera mechanism) that was equipped with numerous gears to simulate positions of astronomical bodies [3]. The sketchbooks of Leonardo da Vinci, dating to the mid 1400s, depict various unique gear mechanisms. Initially, wood was the material of choice for gear manufacture until it was subsequently replaced by cast iron.

A more advanced approach to gear engineering came into being at around 1400 with more comprehensive use of science and mathematics in gear design and the associated mechanisms. The first major investigation into gear design as regards to proving the benefits of the involute curve over a cycloidal was conducted by Philip de la Hire in France and later confirmed by a Swiss mathematician Leonard Euler who was responsible for the law of conjugate action [1]. The industrial revolution in England during the 18th century led to the use of cycloidal gears for clocks, irrigation devices, water mills, and powered machines. Further uses were rapidly developed and explored with the invention of the locomotive, vehicles, and other machines. Gear hobbing and shaping technologies were developed in the early 19th century providing the foundation for fabrication of better quality commercial gears. Various new gear types, materials, and surface treatment techniques were introduced during the 19th century. Further advancement in gear manufacturing, measurement techniques, and testing technologies during the late 19th and in early 20th centuries led the way for significant growth in its application in industry.

1.2 CLASSIFICATION AND GEAR TYPES

A wide range of gear types exist to fulfill various different application requirements. Gears and gear systems are usually classified according to the orientation/arrangement of its associated rotational axes. Gears are therefore classified as parallel-shaft gears, intersecting-shaft gears and nonparallel nonintersecting-shaft gears (refer Table 1.1). The details regarding these three categories and the corresponding gear types are discussed in detail in the following subsections.

1.2.1 Parallel-Shaft Gears

The first and most common class of the gear is where the shaft axes are in the same plane and parallel to one another. The gear teeth may be either cut straight (spur gear) or inclined (helical) and may be of either external or internal configurations. These gears can be either cylindrical or linear-shaped gears, and are used in three main transmission arrangements: External, internal, and rack and pinion.

Spur gears are one of the most extensively used types of parallel-shaft gears. These gears have straight teeth cut parallel to the shaft axis. Engagement by spur gears occurs between two parallel shafts or between a shaft and a rack. The larger of two engaged gears are referred to as the 'gear,' while the smaller is referred to as the 'pinion' regardless of which gear is being driven or acting as the driver. The external configuration of spur gears implies that the driver gear rotates in an opposite direction to the driven gear. A **rack-and-pinion** configuration is a special category of parallel-shaft gears, where transmission occurs by meshing a rack (shaftless linear-shaped gear) and a pinion (cylindrical gear wheel). The rack-and-pinion configuration is extensively used in machine tools and other devices to convert linear motion into rotary motion and vice versa. A **spur rack** is essentially a gear wheel with an infinite radius (Fig. 1.2) that engages a spur gear (pinion) with any number of teeth. An internal spur gear is made with the teeth cut on the inside face of a cylindrical gear which engages with an externally configured gear of matching teeth pattern with both rotating in the same direction (Fig. 1.3). The internal gear is usually referred to as the ring gear or annulus and is often used in planetary gear systems. The best functional performance requires that the diameter of ring gear be at least 1.5 times that of the mating external gear.

TABLE 1.1 The Three Major Categories of Gear Classification and Corresponding Gear Types

Categories of Gears (Based on the Orientation of Gear Shafts)	Types of Gears	Representation	Features, Applications and Methods of Manufacture
Parallel-shaft gears	**Spur gears**		Features: Simple to design and manufacture, highest efficiency, easy to assemble, offer excellent precision, high wear and noisy operation
			Applications: Automotive transmission; Industrial drives; Machine tools; Motors and pumps; Agriculture equipment; Scientific instruments; Electronic devices; Large mills
			Methods of Manufacture: Hobbing, shaping, milling, broaching, casting, extrusion, stamping, powder metallurgy, forging, rolling, grinding, shaving, lapping, honing
	Helical gears		Features: High strength, smooth action, silent operation, offer good precision, axial thrust complications
			Applications: Automotive transmission; High-speed drives and machines; Rolling mills; Robotics; Agricultural equipment
			Methods of Manufacture: Hobbing, milling, shaping, casting, extrusion, stamping, powder metallurgy, forging, rolling, grinding, shaving, honing, lapping
	Rack and pinion		Features: Efficiently convert rotary motion to linear and vice versa; Flexibility
			Applications: Materials handling, linear actuators, power-steering system, machine tools, traveling gantries, robots, positioning systems, stair lifts
			Methods of Manufacture: Milling, shaping, broaching, casting, grinding, shaving, honing

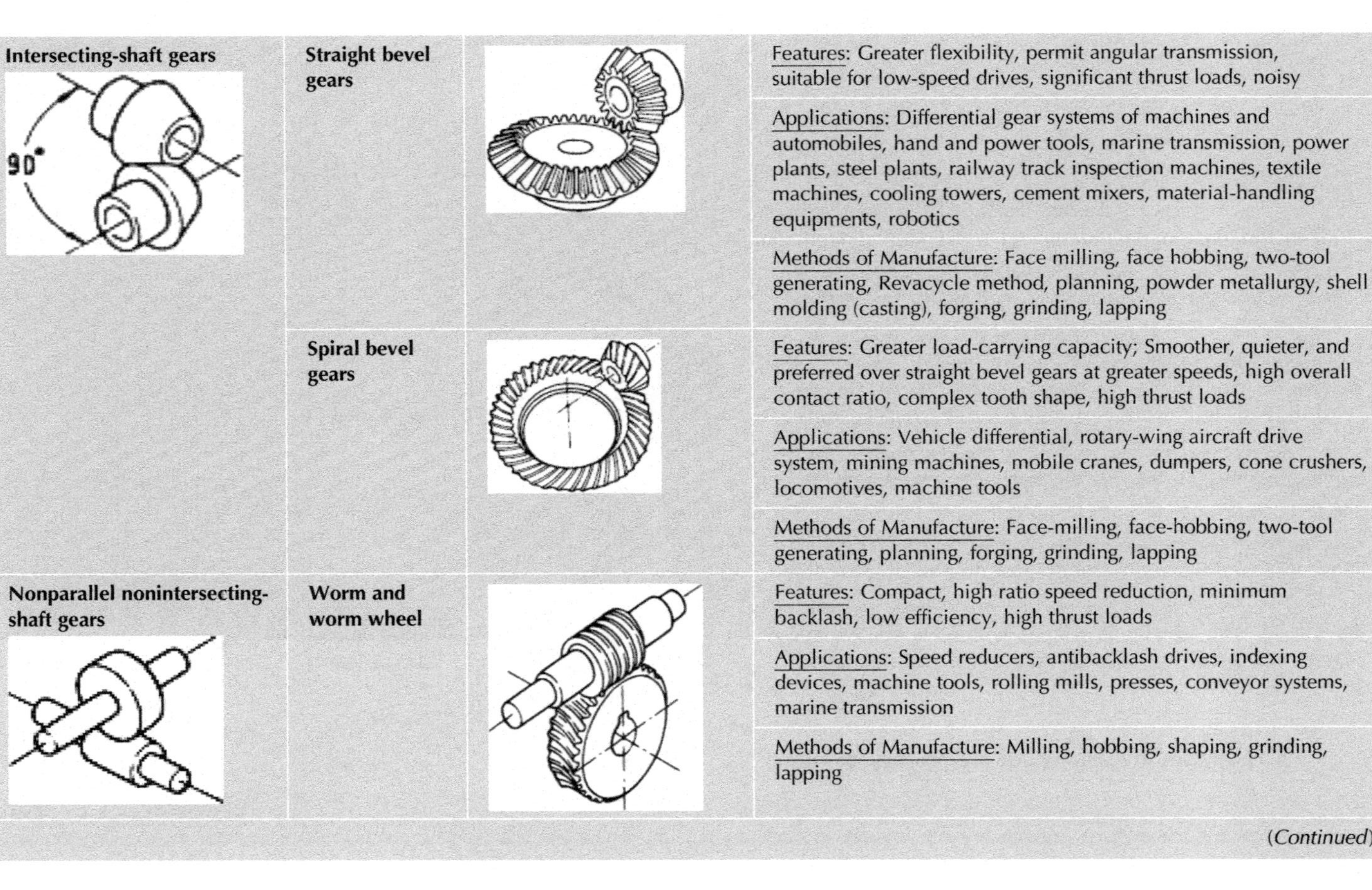

Intersecting-shaft gears	**Straight bevel gears**		Features: Greater flexibility, permit angular transmission, suitable for low-speed drives, significant thrust loads, noisy
			Applications: Differential gear systems of machines and automobiles, hand and power tools, marine transmission, power plants, steel plants, railway track inspection machines, textile machines, cooling towers, cement mixers, material-handling equipments, robotics
			Methods of Manufacture: Face milling, face hobbing, two-tool generating, Revacycle method, planning, powder metallurgy, shell molding (casting), forging, grinding, lapping
	Spiral bevel gears		Features: Greater load-carrying capacity; Smoother, quieter, and preferred over straight bevel gears at greater speeds, high overall contact ratio, complex tooth shape, high thrust loads
			Applications: Vehicle differential, rotary-wing aircraft drive system, mining machines, mobile cranes, dumpers, cone crushers, locomotives, machine tools
			Methods of Manufacture: Face-milling, face-hobbing, two-tool generating, planning, forging, grinding, lapping
Nonparallel nonintersecting-shaft gears	**Worm and worm wheel**		Features: Compact, high ratio speed reduction, minimum backlash, low efficiency, high thrust loads
			Applications: Speed reducers, antibacklash drives, indexing devices, machine tools, rolling mills, presses, conveyor systems, marine transmission
			Methods of Manufacture: Milling, hobbing, shaping, grinding, lapping

(*Continued*)

TABLE 1.1 (Continued)

Categories of Gears (Based on the Orientation of Gear Shafts)	Types of Gears	Representation	Features, Applications and Methods of Manufacture
	Cross-helical (screw) gears	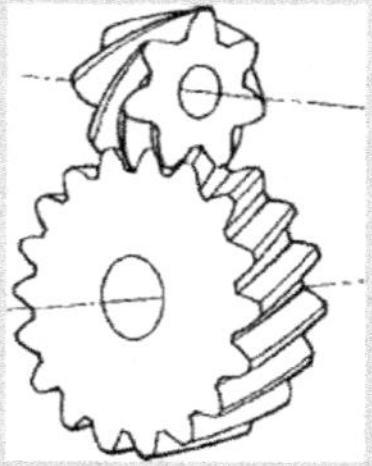	Features: Poor precision rating, allows wide range of speed ratios without changing gear size and center distance, high sliding loads between the teeth
			Applications: Light-load applications; Speed reducers; Agriculture equipment; Tractors; Electric motors, pump drives, substitute for bevel gears
			Methods of Manufacture: Hobbing, shaping, milling, grinding, shaving, lapping, honing
	Hypoid gears	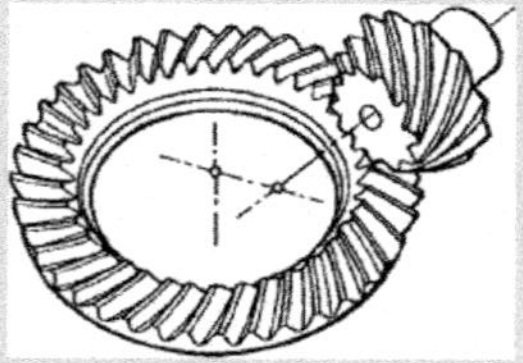	Features: High strength and rigidity, stronger, and quieter than spiral bevel gears, withstand shock-loads, large speed reduction is possible, high reliability, uniform motion, high contact pressure between teeth requires extreme lubrication
			Applications: Industrial machines, automobile differential, speed reducers, agricultural equipment
			Methods of Manufacture: Face hobbing, Face milling, planing, powder metallurgy, grinding, lapping

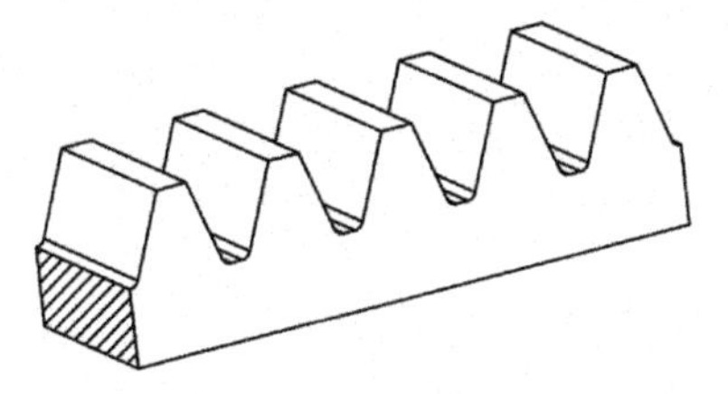

FIGURE 1.2 Spur rack.

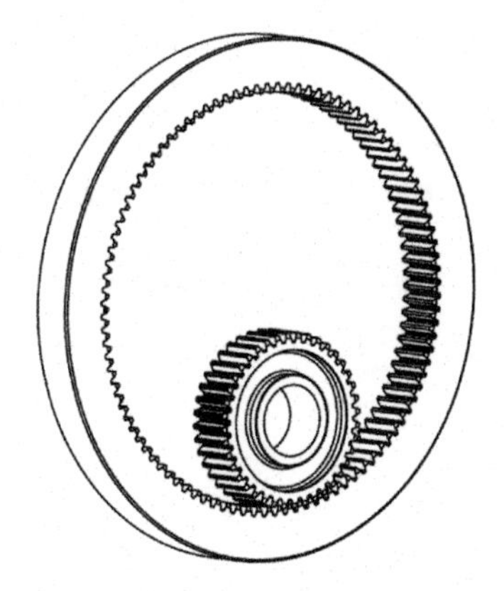

FIGURE 1.3 Internal spur gear.

Because of the straight-tooth nature of spur gears, they are simple in configuration and easy to manufacture. When engaged, spur gears make line contact; whereas during rotation, the contact is mostly of a rolling nature. The mechanical efficiency of spur gears is therefore high. However, these gears suffer from a major drawback in the form of noise and vibration that is the result of the simultaneous contact of teeth upon engagement that results in a continuous shock loading across the entire tooth face. Therefore, spur gears are mostly suitable for low to medium speed applications.

Helical gears are the second most extensively used parallel-shaft gear type. The teeth of these gears are cut in the form of a helix and at an angle to the shaft axis. Typically, helix angles of between 8 to 30° are used with maximum values up to 45° employed in applications where large resultant axial thrust forces are to be avoided [5]. Helix angles may be of either right-hand or left-hand orientation. The teeth of a left-hand gear lean to the left and the teeth of a right-hand gear lean to the right when the gear is placed on a flat surface (as shown in Fig. 1.4A and B). The helix angles of two engaged helical gears must be the same but of opposite orientation for

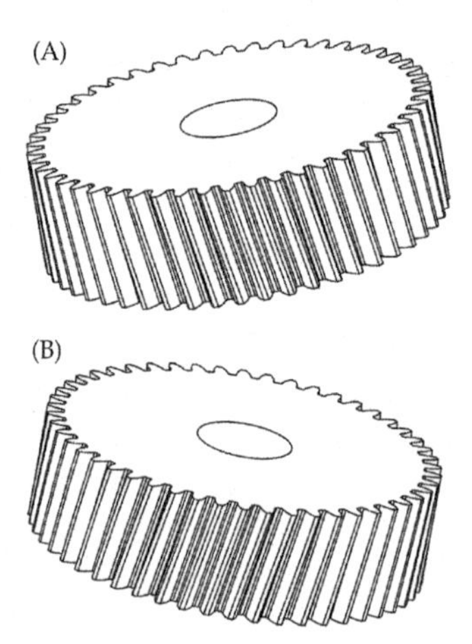

FIGURE 1.4 (A) Left-hand helical gear; (B) right-hand helical gear.

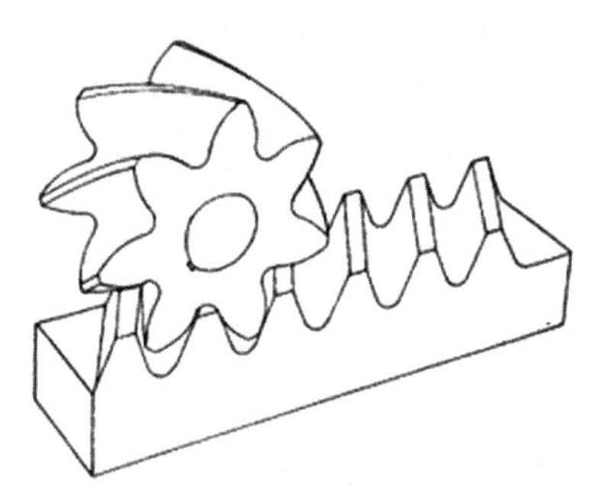

FIGURE 1.5 Helical rack-and-pinion arrangement.

engagement to be possible. Fig. 1.5 illustrates the engagement of a liner-shaped **helical rack** and pinion of same helix angle but of opposite orientation.

Because of the inclined teeth pattern, two or more teeth remain in contact at any given instant during engagement of a pair of helical gears ensuring gradual contact, which provides a smooth and quiet operation. In general, the mechanical strength of helical gears is also higher when compared with spur gears. Despite the advantage of higher strength and therefore greater load-carrying capacity along with lower noise and vibration than an equivalent-size spur gear, the associated thrust force along the rotational axis present major difficulties. It may lead to increased power losses, increased design complications (need for thrust bearings), and reduced system life due to untimely failure.

Double helical gears overcome the problem of thrust loads by counter-balancing them. A double helical gear incorporates two opposite orientated

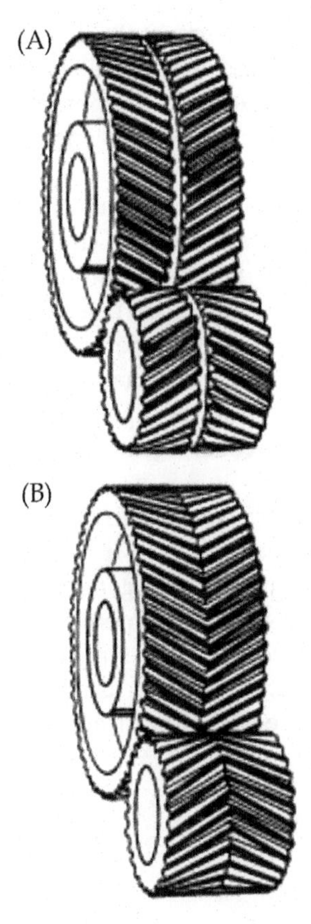

FIGURE 1.6 (A) Double helical gear; (B) herringbone gear.

helical gears of the same helix angle together such that the two opposing thrust forces annihilates one another. The two opposing helixes permit multiple tooth engagement and eliminate end thrust. There are two important configurations for this gear: A double helical gear (Fig. 1.6A), where a small gap exists in between the two opposing helixes and a herringbone gear (Fig. 1.6B), where the opposing helical gears are joined together without this gap.

Internal helical gear arrangements are also possible and comprises of a ring gear with helical teeth cut onto its inside face along with a small external gear (pinion) of same helix angle and orientation.

1.2.2 Intersecting-Shaft Gears

In this gear arrangement, the shaft axes intersect although they are in same plane. The most extensively used configuration is the right angle system, where the two engaged gear shafts are at 90° to each other. This right-angled arrangement is best accomplished by conical-shaped **bevel gears**. There are two basic classes of bevel gears i.e., straight bevel gear and spiral bevel gear. The tapered nature of the bevel gear implies an axial thrust onto the support bearings similar to helical gears.

Straight bevel gears are the simplest to produce and the most widely applied conical gear type. These gears have straight teeth cut along the pitch cone that if extended would intersect with the shaft axis (as shown in Fig. 1.7A and B). Moreover, the teeth are tapered in thickness along the face width and may have either constant or tapered height [6]. The areas of application of straight bevel gears are generally limited to low-speed drives, where vibration and noise may not be significant. These gears are however used for automobile differential gear system and other industrial applications.

Spiral bevel gears are more complex to manufacture because of the spiral (curved) teeth with helical angles (Fig. 1.8). Nonetheless, the curved and oblique teeth ensure gradual engagement with higher contact ratio when engaged and therefore results in smoother and quieter operation than equivalent straight-tooth bevel gears. A typical spiral angle used for these gears is 35° [5]. Spiral bevel gears are usually employed for high-speed applications (typically above 300 m/min) and large speed reduction ratio applications [7].

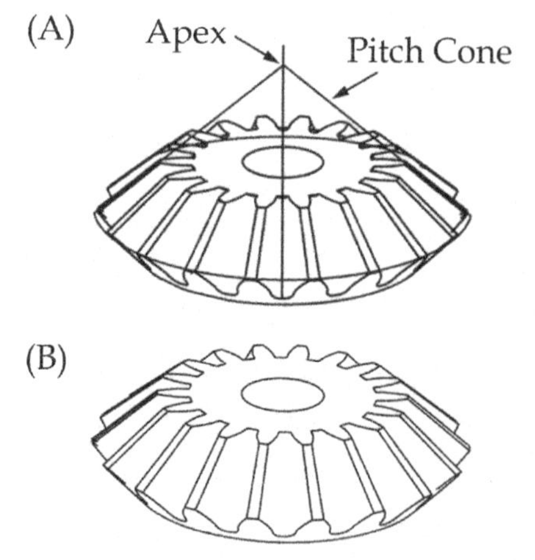

FIGURE 1.7 (A) Direction of teeth cut in straight bevel gear; (B) straight bevel gear.

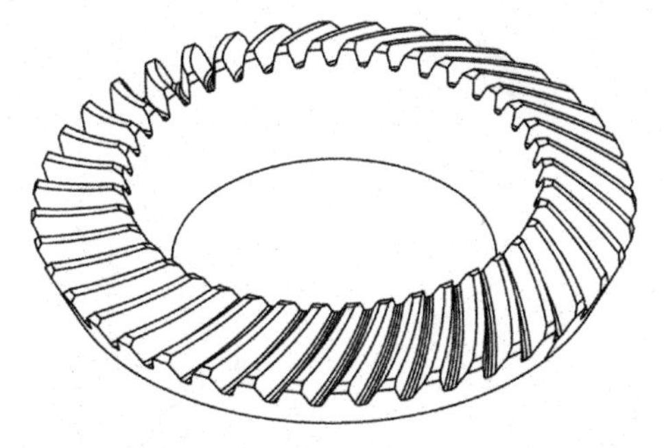

FIGURE 1.8 Spiral bevel gear.

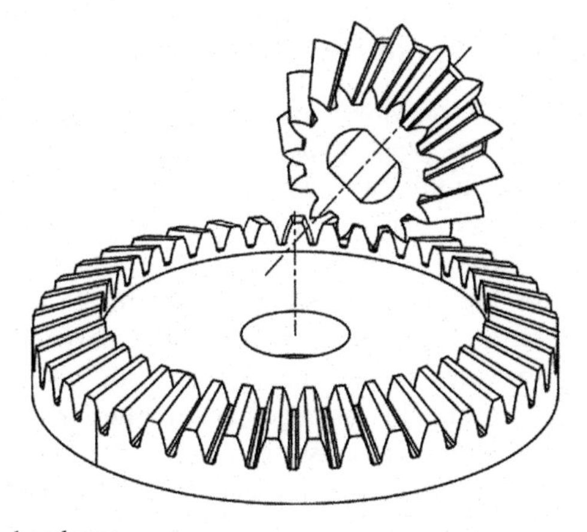

FIGURE 1.9 Zero bevel gear.

Zero bevel gears are another important type of bevel gear with curved teeth with zero spiral angles, which means that the teeth are not oblique (Fig. 1.9). In essence, this implies that the two tooth ends are in the same plane as the gear axis (coplanar). They possess characteristics of both straight and spiral bevel gears. Strength wise, they fall in between straight bevel and spiral gears and are therefore generally employed for medium-load applications. A typical choice of pressure angle for zero bevel gears ranges from 14.5 to 25°.

Miter bevel gears are a special class of bevel gear, where the gear shafts intersect at 90° (each of the two gears has a 45° pitch angle) and both gears have the same number of teeth i.e., the gear ratio is 1:1. Fig. 1.10 depicts a typical miter gear arrangement. Miter gears may be of straight or spiral tooth profile.

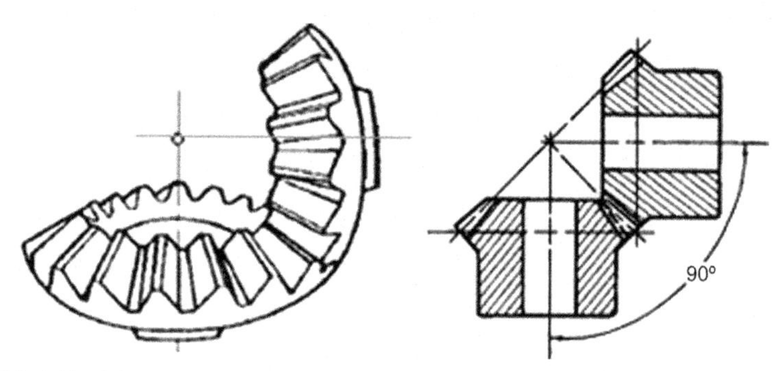

FIGURE 1.10 Miter bevel gear.

1.2.3 Nonparallel Nonintersecting-Shaft Gears

These are noncoplanar gears whose shaft axes can be aligned at any angle between 0 and 90°. Worm gears, hypoid gears, and cross-helical gears are the important gear types in this category.

Worm gears: A worm gearset consists of a worm wheel and a worm whose shafts are placed at a right angle to one other. Worm gears are mainly employed for high-ratio speed reduction in a limited space. The worm resembles a screw thread and meshes with a larger gear referred to as a worm gear (also called worm wheel) with teeth cut at various angles to be driven by the worm. This arrangement implies that large reduction ratios can be obtained with this gearset as a full rotation of the worm only advances the worm gear by the circumference associated with one tooth on the gear wheel. Although the transmission efficiency of worm gears is poor due to significant sliding motion, the screw action of this drive results in quiet and smooth operation. Speed reducers, indexing devices, machine tools, and antireversing gear drives are some of the important applications where worm gears are used.

The **Cylindrical Worm Gear** is the simplest form of worm gear where a straight cylindrical worm engages with a simple helical gear. Two other improved forms of worm gearsets also exist. These are single-enveloping and double-enveloping worm gears which differ as regards to the axial profile of the worm. A **single-enveloping worm gear** is shown in Fig. 1.11A, in which a straight-sided cylindrical worm meshes with a worm wheel which is essentially a throated helical gear and tends to wrap around the worm. This results in greater tooth contact area and thus smoother transmission with high load-carrying capacity. On the other hand, in a **double-enveloping worm gear** both gears are throated and wrap around each other (Fig. 1.11B). This mutual (two-sided) enveloping brings more teeth into contact and provides higher load-carrying capacity as compared with the other worm gear types mentioned above. The shape of the worm for a double-enveloping arrangement is referred to as an "hourglass".

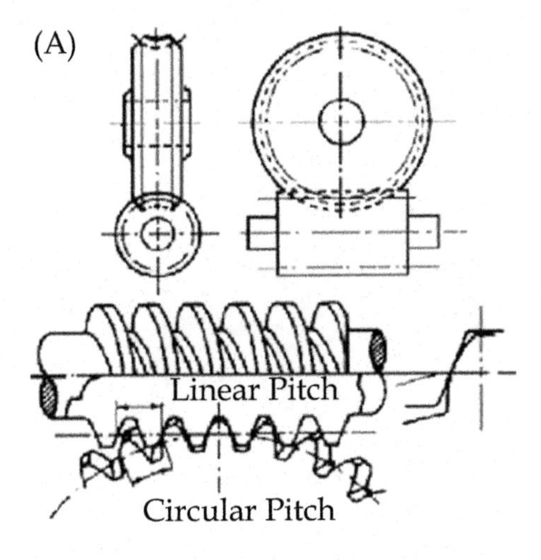

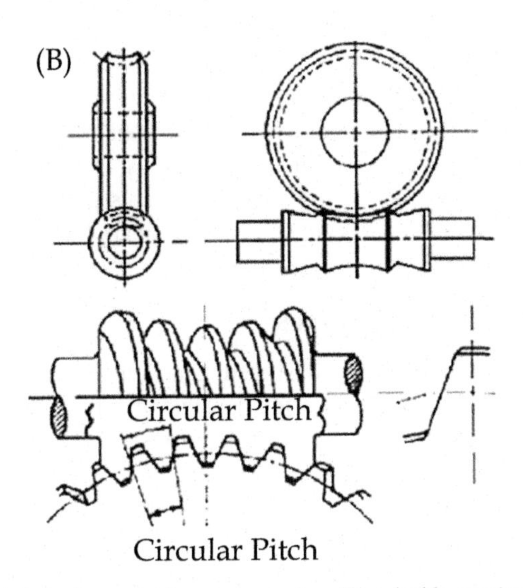

FIGURE 1.11 (A) A single-enveloping worm gearing; (B) a double-enveloping worm gearing.

Hypoid gears are similar to spiral bevel gears except that their pitch surfaces are hyperboloids rather than cones and the pinion axis is somewhat offset from the gear axis i.e., gear-shaft axes do not intersect. The general form of a pair of hypoid gears are shown in Table 1.1. Hypoid gears exhibit improved smoothness and lower noise when compared with spiral bevel gears due to the higher overall contact ratio. These gears find extensive applications in differential gear units in rear axles of automobiles with rear-wheel drives (Fig. 1.12).

Crossed-helical gears or **screw gears** resemble a gearing arrangement where two helical gears of either the same or opposite orientation meshes with

FIGURE 1.12 A typical bevel gear differential unit with straight bevel and hypoid gears.

their axes crossed at an angle. When installed in a crossed arrangement, they are also referred to as screw gears. Both of the engaged helical gears must have the same normal pressure angle and diametral pitch but may have different helix angles and orientations. Because of the sliding action and limited tooth contact area (point contact), the application domain for screw gears is limited to light loads only. These are not suitable for high-power transmission requirements. A typical arrangement of crossed-helical gearing is depicted in Table 1.1. Automatic machines that require intricate movements, pump drives, electrical motors, and cam-shafts of engines are some common applications of these gears.

1.2.4 Some Special Gear Types

Noncircular gears are specially designed gears used to transmit a unique motion and/or to convert speed in a nonconstant manner between two parallel axes shafts. These gears find applications in flow meters, textile machines, high-torque hydraulic engines, Geneva-mechanism, printing press equipment, pumps, packaging machines, potentiometers, conveyors, windshield wipers, robotic-mechanisms etc. The main purposes of these gears are to improve the function, versatility, and simplicity of the mechanical operations. Noncircular gears such as triangular, rectangular, square, elliptical, and oval-shaped gears are employed (as shown in Fig. 1.13) to perform a variety of mechanical functions, including complex displacement and velocity changes. Noncircular gears are also used as a viable substitute for cam-follower mechanism as they may provide a more compact and more accurate solution. Conversion of constant input speed into a variable output speed and stop-and-dwell motion requirements are efficiently handled by these gears. To fulfill these unique requirements noncircular gears of any particular shape can be assembled or meshed with gears of any other shape. Noncircular planetary gears are also used as effective motion convertors.

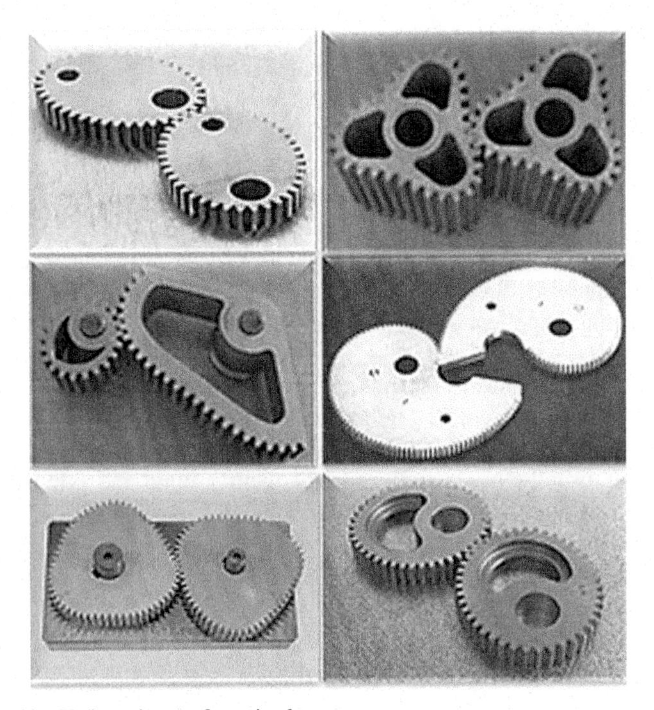

FIGURE 1.13 Various shapes of noncircular gears.

FIGURE 1.14 Close-up of a sector gear.

A **Sector gear** is a gear that has teeth cut along a section of its circumference only. The remaining portion of the gear is smooth without any teeth (Fig. 1.14). This arrangement may save space, material, and manufacturing costs. Sector gears are used in applications when less than 360° of rotation is required. These gears are typically used in applications such as valve actuators that only require 90° to open or close aircraft radar that scans through a limited angular range and X-ray machines etc. Spur, helical, bevel, and/or even worm gears may be used as sector gears and may be fabricated from most materials commonly used in gear manufacture.

A **Ratchet wheel** is a circular wheel provided with saw-shaped teeth (Fig. 1.15) that allows continuous rotational or intermittent motion in one direction but limits it in the other. Ratchets are widely used in machinery and tools. When used in a **pawl-and-ratchet mechanism**, as the ratchet wheel turns, the pawl which is a spring-loaded, finger-shaped element falls into the tooth cavity that then effectively "locks" the gear wheel from rotation in the opposite direction. A ratchet wheel can also be used to arrest motion. Common applications of this mechanism include mechanical clocks, ratchet spanners, winders, and jacks.

Splines are mechanical elements that have ridges or teeth on a shaft and that mesh with grooves cut in a hub in order to transfer torque and/or motion along the same axis. These are used in many mechanical drive systems. A splined shaft usually has equally spaced teeth around the circumference, which are most often parallel to the shaft's axis of rotation. These teeth can be straight sided, at an included angle, or of involute form. The externally splined shaft mates with an internal spline that has slots or spaces formed to

FIGURE 1.15 A ratchet wheel.

FIGURE 1.16 A typical splined connector.

FIGURE 1.17 A flange-type gear coupling.

the reverse of the shaft's teeth. In other words, splines are always used as a combination of one external and one internal element. Fig. 1.16 shows a typical splined connector.

Splines are generally used in three types of applications: For coupling shafts when significant torque are to be transmitted without slippage; for transmitting power to gears, pulleys, and other rotating elements that are either fixed or mounted such that linear motion is possible (slide); and for attaching parts that may require removal for indexing or adjustment in angular position [8]. Hobbing, shaping, broaching, and cold rolling are the methods used for manufacturing of splines.

Gear coupling is a mechanical device used to transmit torque between two noncolinear shafts. It consists of a flexible joint (which comprises of two hubs and a sleeve) fixed to each shaft (as shown in Fig. 1.17). Both the hubs mounted on the shafts have external gear teeth that engage with internal teeth cut into a sleeve which is fitted over both hubs. The primary purpose of couplings is to join two elements of rotating equipment while permitting a certain degree of misalignment or linear end movement or both. Fig. 1.17 shows a typical flange-type gear coupling. Gear couplings are the most widely used flexible-type shaft coupling.

1.3 GEAR TERMINOLOGY

Gears are generally specified by their types (e.g., spur, bevel, spiral, etc.), size or dimensions, geometry, materials, and special features (if any). This section discusses some common and general gear terms, some special terms with reference to the dimensions and the geometry and meshing conditions.

Involute and Cycloidal Profile

The gear tooth profile is mainly based on two engineering curves i.e., involute curve and cycloidal curve. An involute of a circle is essentially a plane curve generated by a point on a tangent that rolls on the circle without slipping. In other words, an involute curve is developed by tracing a point on a cord as it unwinds from a circle which is the base circle. In cycloidal gears, the face section of a gear tooth profile is constructed by an epicycloid which is the curve traced by a point on the circumference of a circle that rolls without slipping on the outside of a fixed circle i.e., pitch circle; and the flank section is constructed by a hypocycloid, where the circle rolls inside of the pitch circle.

Cycloidal gears are specifically used in clocks and watches, while involute gears have widespread uses such as in machine tools, vehicle gear boxes, robotics, home appliances, scientific instruments etc.

Involute gears have certain advantages over cycloidal gears i.e., easy to manufacture as face and flank are both generated by a single curve; the center distance for a pair of involute gears can be varied within limits without changing the velocity ratio; and the pressure angle remains constant from the start to the end of the engagement. Involute gears does however have the disadvantage that interference may occur in which the root of one tooth undercuts the tip of another during meshing; whereas with cycloidal gears interference does not occur. In general, cycloidal gears are stronger than involute gears due to their wider flanks.

The following terms are commonly applied to the various classes of gears (refer Fig. 1.18):

Pitch circle or reference circle: It is an imaginary circle which by pure rolling action, would give the same motion as the actual gear produces. This imaginary circle also passes through the center of each tooth on the gear and has the gear axis as its center.

Pitch circle diameter or reference circle diameter (*d*): It is the diameter of the pitch circle and an important term to specify the size of the gear.

Module (*m*): It is the ratio of the pitch circle (reference) diameter of a gear to the number of teeth. It defines the size of a tooth in the metric system and is usually denoted by m. The unit of modules should be in millimeters (mm).

$$m = \frac{d}{N},$$

where N is the number of teeth

Preferred module values are 0.5, 0.8, 1, 1.25, 1.5, 2.5, 3, 4, 5, 6, 8, 10, 12, 16, 20, 25, 32, 40, and 50. Fig. 1.19 depicts the relative size of teeth machined in a rack with module values ranging from 0.5 to 6 mm.

Diametral pitch (*DP*): It is the number of teeth per inch of pitch diameter. In English system, it is a measure of tooth size. The higher the value of diametral pitch, the lower is the tooth size. The diametral pitch usually varies between 200 to 1 [9].

Transformation from diametral pitch to module

$$m = \frac{25.4}{DP}$$

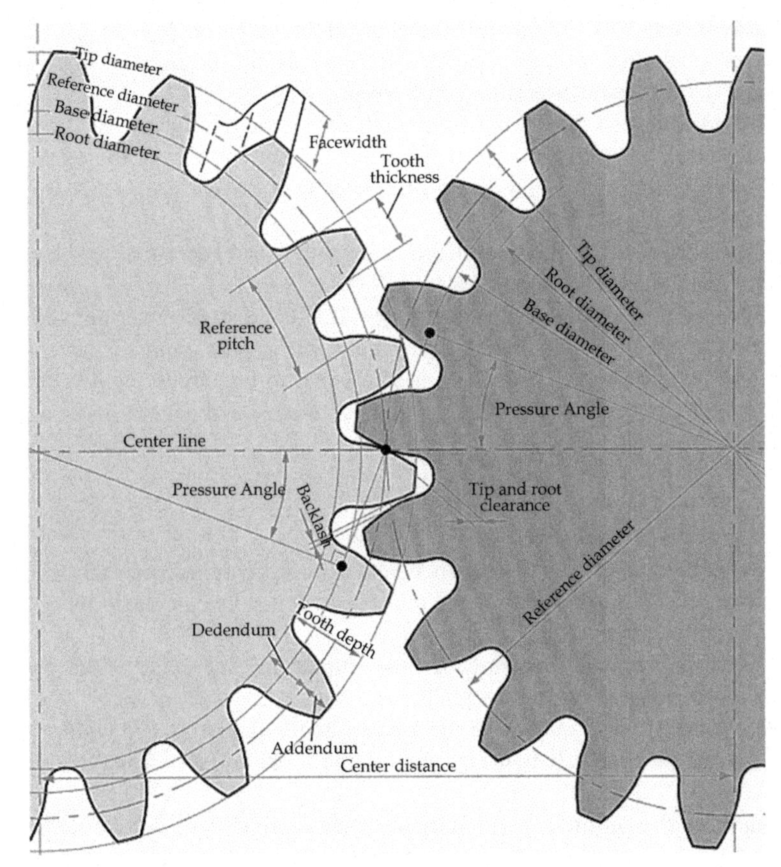

FIGURE 1.18 Schematic showing principal terminology of gears.

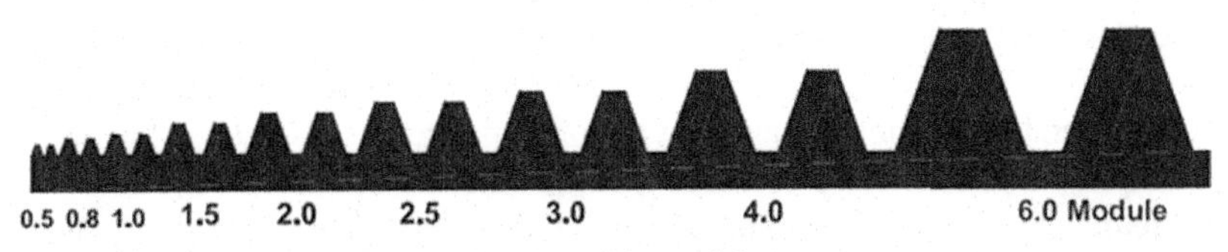

FIGURE 1.19 Gear tooth size as a function of the module.

Base diameter: It is the diameter of the base circle where the involute portion of a tooth profile starts.

Root diameter: It is the diameter of the circle that contains the roots or bottom land of the tooth spaces. In other words, it is the diameter at the base/bottom of the tooth space.

Tip diameter: It is the diameter of the circle that contains the tops of the teeth or top land of external gears. It is also called the outside diameter.

Addendum: It is the height by which a tooth projects beyond the pitch circle or pitch line. It can also be described as the portion of the tooth between the tip diameter and the pitch circle.

Dedendum: It is the depth of the tooth below the pitch circle or the portion of the tooth between the pitch circle and the bottom land of the tooth.

Working depth: It is the depth of engagement of two gears, that is, the sum of their addendums.

Whole depth: It is the total depth of a tooth space and equal to the sum of addendum and dedendum.

Circular pitch or reference pitch (p): It is the distance measured along the pitch circle or pitch line from any point on a gear tooth to the corresponding point on the next tooth. It is also equal to the circumference of the pitch circle divided by the total number of teeth on the gear and consequently formulated as:

$$p = \pi m,$$

where m is the gear module.

Center distance: It is the distance between the center of the shaft of one gear to the center of the shaft of the other gear. It can be calculated by dividing the sum of the pitch diameters by two.

Tooth thickness: It is also called the chordal thickness and is the width of the tooth measured along the pitch circle.

Top land: It is the surface of the top of a tooth along the face width.

Face of the tooth: It is the portion of the tooth surface above the pitch surface.

Flank of the tooth: It is the portion of the tooth surface below the pitch surface.

Face width: It is the length of a tooth in an axial plane and measured parallel to the gear axis.

Pressure angle (ϕ): Pressure angle is the angle at which the pressure from the tooth of one gear is passed on to the tooth of another gear. Geometrically, it is the angle between the common normal between two gear teeth at the point of contact and the common tangent at the pitch point. It is also called the angle of obliquity. The standard pressure angles are 14½ and 20°.

Clearance: It is the amount by which the dedendum in a given gear exceeds the addendum of its mating gear.

Backlash: It can be defined as the amount by which the width of a tooth space exceeds the thickness of the engaging tooth on the pitch circles.

Contact ratio: To assure continuous smooth tooth action, as one pair of teeth disengages a subsequent pair of teeth must already have come into engagement. It is desirable to have as much overlap as possible. Contact ratio is the measure of this overlapping and simply indicates the number of tooth pairs engaged at any given time. Higher contact ratios are always desirable and if it becomes less than 1.0, the gears disengage.

1.3.1 Standard Gear Tooth Proportions

Four systems of gear tooth proportions are used to achieve engineering interchangeability between gears belonging to the same category and therefore having the same pitch and pressure angle. These four systems are:

1. 14½° Composite system;
2. 14½° Full-depth involute system;
3. 20° Full-depth involute system; and
4. 20° Stub involute system.

The 14½° composite system has a cycloidal tooth profile for both top and bottom regions with an involute profile in the center. This system is used for general purpose gears. As the name implies, the 14½° full-depth involute system has an involute shaped tooth profile. The 20° full-depth involute system has a wider base and is therefore used in applications that require increased strength. The 20° stub involute system is used for applications that requires maximum strength to sustain exceptionally high loads. These systems imply specific gear geometry descriptors, including the addendum, dedendum, tooth thickness, and depth. The standard proportions for these (as a function of the module, *m*) are presented in Table 1.2.

TABLE 1.2 Standard Proportions of Systems for Spur Gear [5,6]

Term	14½° Composite system or Full-Depth Involute System (*m*)	20° Full-Depth Involute System (*m*)	20° Stub Involute System (*m*)
Addendum	1	1	0.8
Minimum dedendum	1.25	1.25	1
Working depth	2	2	1.60
Minimum whole depth	2.25	2.25	1.80
Basic tooth thickness	1.5708	1.5708	1.5708
Minimum clearance	0.25	0.25	0.2
Fillet radius at root	0.4	0.4	0.4

1.4 GEAR MATERIALS

The selection of a suitable material for any gear type is influenced by a number of technical and commercial factors which include transmission requirements in terms of power, speed, and torque; working environment, i.e., temperature, vibration, and chemical conditions; ease of manufacture; cost of processing and manufacture etc. An appropriate material choice is required for a gear to achieve its intended performance with adequate reliability throughout its intended service life. High strength, toughness, and resistance to fatigue and wear are some important and desirable characteristics for gear materials. Gears are extensively manufactured in both metallic and nonmetallic materials. Cast iron, plain carbon steels, and special alloy steels are important ferrous gear materials used for heavy-duty applications, whereas brass, aluminum, and bronze are frequently used nonferrous materials for medium to light-duty applications. Nonmetallic materials such as plastics are generally used for light-duty service applications in noise sensitive environments. Both the material properties and application requirements should be considered for appropriate final material selection. Table 1.3 presents special features and typical applications of some important gear materials which are also discussed in greater detail in the following sections.

1.4.1 Ferrous Metals and Alloys

1.4.1.1 Cast Iron

Properties such as low cost, good machinability, excellent noise-damping characteristics, and superior performance under dynamic conditions make cast iron eminently suitable as a gear material. The three basic grades of cast iron used for gears are gray, ductile, and malleable cast iron. Due to low shock resistance, gray cast iron cannot be used in gears that are subjected to shock load applications. Ductile iron has good impact strength and sufficient ductility. The fatigue strength of ductile iron is on par with steel of equal hardness [10]. The bending strength of cast iron are between 34 and 172 MPa while the surface fatigue strength ranges between 345 and 793 MPa [7]. Cast iron gears have good corrosion resistance and are generally quieter in operation than steel gears. These gears are also a good replacement for bronze worm gear drives due to their low cost and good sliding properties. Cast iron is commonly used in applications, including equipment and machinery used for mining, earthmoving, agriculture, construction, and machine tools.

1.4.1.2 Steel

Because of its high strength-to-weight ratio and relatively low cost, various different grades of steel are regularly used in applications requiring medium to heavy-duty power transmission, high-strength applications, and precision requirements. Both plain carbon and alloy steels are used in various different applications even though the latter may be difficult to machine. Typical commercial grades of plain carbon steels used are mild steel (AISI 1020, EN3), AISI 1040 (EN 8) and EN 9. These are used in applications requiring

TABLE 1.3 Summary of Properties and Applications of Some Important Gear Materials [6,9,10]

			Properties	Applications
Cast iron	Gray iron		Good machinability, sound dampening properties, good resistance to wear, low impact strength	Large-size mill gears; moderate power-rating applications; low shock applications; and machine tools
	Ductile iron		Fair to good machinability, sound dampening properties, better impact and fatigue strength than gray iron	Transportation; railroad and military vehicles; girth gears for mills
Plain carbon steels	Carburizing gear steels	Low-carbon steels (1010, 1015, 1020, 1021, 1022, 1025)	Excellent machinability, good combination of strength and ductility, heat treatable, can be case carburized	Low to medium duty applications
	Through-hardening gear steels	Medium-carbon steel (1035, 1040, 1045)	Good machinability	Moderate to high power-ratings application
		High-carbon steel (1060)	High strength and durability	High-power rating applications
Alloy steels	Carburizing steel	Nickel–chrome–molybdenum carburizing steel (SAE8620)	Good wear characteristics/ high wear resistance	Automotive transmissions; farm machineries; earth movers
		20MnCr5 (SAE5120)	Case-hardening imparts hard case with good wearing properties and tough core	Automobile gear boxes; heavy-duty transmission gears; hoisting; and cranes
	Through-hardening gear steels	Chrome–molybdenum alloy steel (4140)	High toughness, good torsional strength, good fatigue strength	Differential systems of automobiles; and tractors

(*Continued*)

TABLE 1.3 (Continued)

			Properties	Applications
		Nickel–chrome–molybdenum alloy steel (4340)	Can be hardened easily, toughness and high strength in heat-treated condition, good fatigue strength	Industrial drives; mining equipment, paper mills; steel mills
Stainless steels	300 series	303, 304, 316	Fair machinability, high strength, high corrosion resistance, nonmagnetic, cannot be heat treated	Precision applications; low-duty applications; antibacklash gears; juice extractors; fishing reels; gear reducers; underwater applications; gear pumps; crushers; medical instruments; and microgears
	400 series	416, 440 C	Fair machinability, heat treatable, nonmagnetic, high resistance to corrosion, good hardness and wear resistance	Low-to-medium duty applications; precision applications; and miniaturized devices
	Precipitation hardening	17-4PH, 17-7PH	High strength, moderate corrosion resistance, high-fracture toughness	Precision applications and microgears
Brass alloys	Free-cutting brass, yellow brass (die-cast alloy), naval brass		Good machinability, light weight, good corrosion resistance, good wear resistance	Low-duty applications; miniature motors; hydroservice applications; and precision motion transmission applications
Bronze alloys	Phosphor bronze		High toughness and hardness, high fatigue resistance, high wear resistance, good machinability, good corrosion resistance, and nonmagnetic	Used to make gears operate in mesh with steel worm gears for medium-duty applications
	Manganese bronze			
	Aluminum bronze			
	Silicon bronze			Low-duty applications

Aluminum alloys		2024, 6061, 7075	Light weight, excellent machinability, corrosion resistant	Extreme low-duty applications and instrument gears
Magnesium die-cast alloys		AZ91B, AZ91A, AM60, AS41	Light weight, corrosion resistant	Light-load applications
Zinc die-cast alloys			Good impact strength	Light-loading and low-temperature applications; and small-gear trains
Plastics	Thermoplastics	Nylons, acetals (derlin), polysters, polycarbonates, polytetrafluoroethylene (teflon)	Good corrosion resistance, low cost, and silent operation; thermoplastics have greater toughness than thermosetting plastics	Light-load and precision applications; instruments; printers; electronic devices; appliances; and clocks
	Thermosets	Polyurethanes, polyamides, phenolics, and laminated phenolics,		
Ceramics	Alumina		Good corrosion resistance and can withstand high-temperature conditions	Microgears for watches; instruments; microgearboxes; microactuators; and microreducers
	Zirconia			

medium to high toughness and strength requirements. Low and medium-carbon steels are good choices for gear stamping.

Plain carbon and alloy steels are generally heat treatable which implies that certain desirable mechanical characteristics may be imparted on the gears by an appropriate heat treatment. Hardenability, which is primarily a function of the alloy content, may be an important aspect to consider when selecting a gear steel. Surface (case) hardening and through hardening are two important heat treatment operations that may dramatically improve the strength, toughness, endurance limit, and shock resistance of steels. Through-hardened low-alloy steels are suitable for medium duty and moderate operating conditions. Heavy-duty and/or severe operating conditions usually necessitate case-hardened high-alloy steels. Typically, case-hardened gears are produced in a two-stage process (carburizing) by initially enriching the carbon content of the surface locally (up to 0.85%). This is followed by an appropriate quenching and tempering process that produces a hardened case (up to 64 HRC) of required thickness. The core remains largely unchanged and retains its inherent ductility and toughness. Localized quenching and tempering of appropriate steel (containing sufficient carbon usually more than 0.4%) may produce similar results (selective hardening). Elements other than carbon, such as nitrogen and boron, may also be used albeit in a slightly different technique. Generally, case-hardened gears have improved fatigue life and wear resistance. Carburizing, nitriding, laser hardening, and induction hardening are the most important case-hardening techniques employed for steel gears. Typical case depths are between 0.075 to 8.25 mm [10]. Case-hardened gears can withstand higher loads than through-hardened gears, whereas the latter are usually quieter and less expensive [10]. A wide range of steels that are eminently case hardenable exist. These include carburizing steels such as AISI 1015, 1018, 1020, 1022, 1025, 1117, 1118, 4020, 4026, 8720, 9310 etc.; and nitriding steels namely AISI 4140, 4340, 6140, 8740, and nitralloy. Nitralloy N and type 135 are suitable materials for highly stressed heavy-duty gears.

20MnCr5, 16MnCr5, and SAE 8620 are the most important case-hardening steels preferred for automobile gears. In general, alloy steels are preferred over plain carbon steels because of the increased hardness and effective depth of the case for a similar carbon content, finer grain size, lower distortion, increased toughness, and improved wear resistance.

Stainless steels (SS) are important alloy steels eminently suitable for gears exposed to high temperature and corrosive environments such as equipment and machines used in chemical, petrochemical, and the food and beverage-processing industries. Steel is usually designated "stainless" if it contains more that 12% chromium. The chromium reacts with oxygen in the air to form a homogeneous passive oxide layer that protects the core material against corrosion. Austenitic (types 303, 304, and 316), ferritic (type 430), martensitic (type 440 C), and precipitation hardening (17-4PH and 17-7PH) are the major stainless steel types [10]. The austenitic grades are popular for use in extreme environments because of their superior corrosion resistance. They are nonmagnetic, nonheat treatable, and generally difficult to machine. Type 304 (18Cr-8Ni) is one of the most widely used stainless steels for gears employed in extreme corrosion conditions.

Various types of stainless steel namely 303, 304, 316 L, 420, 440, and 17-4 PH steels are also extensively used for manufacturing of high-accuracy miniature gears employed in precision and high-torque transfer applications such as microharmonic drives, robot mechanisms, micromotors, micropumps, speed reducers, medical instruments, and electronic equipment. Types 303 and 17-4 PH are extensively used to fabricate ratchet wheels as used in pawl-and-ratchet mechanisms.

Gears manufactured by the casting process from **casting steels** are used for service conditions involving multidirectional loading in machines such as large mills, crane wheels, and wind turbine gearbox components. Common through-hardening cast steels include AISI grades 4135, 8630, 8640, and 4340, whereas AISI grades 1020, 8620, and 4320 are some important case-hardening cast steels.

Sintered steels are used to manufacture gears by the powder-pressing process. Iron–copper and iron–nickel steels are the two most important sintered steels available in powder form and used to manufacture helical, spur, and bevel gears that are exposed to high-strength applications. Powders of Type 316 L stainless steel and a powder mixture of nickel steel and bronze are also frequently used to manufacture gears by powder-pressing technique.

1.4.2 Nonferrous Metals and Alloys

A wide range of nonferrous metals such as copper, brass, bronze, aluminum, and magnesium are used to manufacture machined, die-cast, and formed gears. Depending on the alloy, they may have good manufacturability, low density, and good corrosion resistance while being nonmagnetic.

1.4.2.1 Copper Alloys

Bronze (copper–tin alloy) and brass (copper–zinc alloy) are the two most extensively used copper alloys. Several types of bronze and brass alloys are used in gearing. **Bronze** alloys are mostly used in worm gear applications, where low-sliding friction, wear resistance, and high-reduction ratios are the prime requirements. The four main types of bronze alloys used are aluminum bronze, manganese bronze, phosphorous bronze, and silicon bronze. All bronze alloys possess good machinability, wear and corrosion resistance, and have yield strengths ranging between 138–414 MPa [7].

A wide variety of **brass** that includes free-cutting brass, yellow brass, and naval brass are used in machined, die-cast, and formed gears of different shapes and size. Die-casting of brass is usually more expensive when compared with the other nonferrous metals. Brass gears are light weight and corrosion resistant. Brass alloys are one of the best choices for manufacturing miniature gears through various conventional and advanced processes for light-load applications as found in precision instruments, miniaturized products, and slow-speed machines.

Zinc and zinc–aluminum alloys are preferentially used to die-cast gears of good finish and accuracy at low cost. **Zn–22Al** alloy in powder form is a good choice to manufacture microgears by extrusion and hot-embossing methods.

1.4.2.2 Aluminum Alloy

Gears made from Al-alloys have the advantage of light weight combined with moderate strength. They are also corrosion resistant, easy to machine, and provide a good surface finish. A major disadvantage of aluminum is its large coefficient of thermal expansion compared to steel. High-strength wrought aluminum alloys (2024, 6061, and 7075) are used for machined gears and aluminum silicon alloys (A360, 383, 384, and 413) are used as die-cast alloys.

Magnesium alloys including ASTM AZ91A; AZ91B; AM60; AS41 are used in light weight die-casted gears for low load applications.

1.4.3 Nonmetals

Plastics including nylon, polyacetal, polyamide, polycarbonate, polyurethane, phenolic laminates etc. are extensively used to manufacture gears. Plastic gears offer light weight, smooth and quiet operation, wear and corrosion resistance, and may be manufactured at relatively low cost. In certain applications, they may be economic substitutes for metallic gears as they are easier to manufacture, require minimum or no finishing, are able to run with minimal or no lubrication, and generally have extended life spans. Hobbing, injection molding, and rapid prototyping techniques are the most popular techniques to manufacture plastic gears. Plastics are specifically preferred for miniature gear manufacture requiring precision and quite operation. In many instances, they may perform markedly better than their metallic equivalents.

Two broad categories of plastic materials are available for gear manufacture i.e., thermosets (polyurethanes, polyamides, phenolics, and laminated phenolics) and thermoplastics (nylons, acetals, polyesters, polycarbonates, fluoropolymers such as polytetrafluoroethylene etc.). Polyamides and polycarbonates are extensively used in selective laser sintering, stereolithography and 3D printing-type additive layer manufacturing techniques for rapid prototyping of gears.

Additives such as mica, carbon powders, kevlar, glass beads and fibers, graphite, ethylene vinyl acetate, acrylics etc. are introduced into plastics with the aim to improve their performance. Plastic gears are used in various different applications, including printers, computer memory devices, robots, toys, electronic devices, micromotors, clocks, small power tools, appliances, speedometers, automotive actuators, medical instruments etc.

Advancements in materials engineering have extended the use of plastic gears into the high-speed and high-torque transmission domains for certain applications.

Ceramics such as zirconia powder (ZrO_2), alumina (Al_2O_3) etc. are the key materials for precise microgears formed by metal powder injection molding techniques and are widely used in watches, microplanetary mechanisms, and microgearboxes. Ceramic gears are viable substitutes for plastic gears for high-temperature conditions and corrosive environments that require high mechanical strength.

The LIGA (a German acronym for Lithographie, Galvanoformung, and Abformung; lithography, electroplating and molding) process uses ceramics, electroplated metals, and photosensitive resins reinforced with ceramic nanoparticles, polymers, and silicon-based materials for fabrication of micro and nanogears that find applications in NEMS-MEMS, actuators, microsystems, and medical devices [11].

1.5 GEAR MANUFACTURE

Conventional metallic gear manufacture usually comprises of several sequential operations depending upon the gear type, material, and desired quality. These operations include preparation of the blank i.e., preforming the blank by casting or forging, heat treatment of the preformed blank (if required), shaping the blank to the required dimensions by machining, producing the teeth in gear blank, heat treatment of the gear (teeth), and finally finishing of the teeth, if required.

In case of large gears, the blank is produced by casting or forging, whereas medium-sized and small gears are produced by cutting and machining a cylindrical billet of the selected gear material to the required size. Sand casting is used for large cast iron gear blanks; while centrifugal casting is frequently used to form the blank for worm gears in cast iron, bronze, or steel. The preformed blank is subjected to a machining operation to produce the required shape and size and removal of excess material from its surface (if any) before teeth machining commences. Other casting techniques such as shell molding, die-casting, metal mold casting, and sometimes even sand casting may also be used to create a preform that may include gear teeth also instead of forming the blank only.

In general, "gear manufacturing" refers to the process required up to the production of the teeth, whereas "gear finishing" refers to the finishing and refinement of the teeth to final requirements.

Certain gear manufacturing techniques may not require blank preparation. Methods other than machining such as die-casting, powder metallurgy, and injection molding may shape the gear teeth close to the final required geometry and dimensions without requiring separate blanks.

A wide range of gear manufacturing and finishing processes exists, including both conventional and advanced types to produce gears of various types, shapes, sizes, and materials.

1.5.1 Conventional Gear Manufacturing

The conventional processes of producing gear teeth are grouped into the following major classes:

I Subtractive or material removal processes

Techniques belong to this group can be classified as:

i Form-cutting processes

Gear cutting by milling
Gear cutting by broaching
Gear cutting by shaper

ii Generative processes

Gear generation with hob cutter: Gear hobbing
Gear generation with rotary cutters: Gear shaping
Gear generation by planing
Generative processes for conical gears

II Forming processes

Stamping and fine blanking
Extrusion and cold drawing
Gear rolling
Forging

III Additive processes

Gear casting
Powder metallurgy
Injection molding for plastic gears

The first and one of the most important gear manufacturing techniques for cutting teeth of required geometry in prepared gear blanks is the **subtractive or material removal process**. All these material removal process types make use of an appropriately shaped tool, according to the gear geometry, to cut the required geometry into the blanks. These processes can further be subdivided in two main groups, namely form-cutting and generative processes.

Form-cutting processes are those where the teeth profiles are obtained as a replica of the geometry of the cutting tool (edge) e.g., milling, broaching, and teeth cutting on shaper.

The **generative process** produces gear teeth as the result obtained (generated) of relative motion between the gear blank and the cutting tool with the cutting tool being of the same geometry as the teeth to be cut. Generally, a cutter reciprocates and/or rotates against a rotating gear blank to cut the teeth. Hobbing, shaping, planning, and other conical gear generative processes such as face-mill cutting, face-hob cutting, Revacycle process, two-tool generators, planning generators etc. fall under this category. Subsequent chapters of this book present a more detailed description of these material removal type gear-manufacturing processes.

A number of forming methods such as stamping and fine blanking, extrusion and cold drawing, and rolling and forging are used to make gears either by forming a complete gear from the raw material or by forming teeth in the blank using dies of appropriate geometry.

Another important class of gear manufacturing is **additive processes** which includes various casting methods i.e., sand casting, die-casting, investment casting etc.; powder metallurgy to make gears from metal powders, injection molding for plastic gears.

The aforementioned conventional processes used to manufacture cylindrical gears are discussed in Chapter 2, Conventional Manufacturing of Cylindrical Gears, whereas manufacturing of conical and noncircular gears is discussed in Chapter 3, Manufacturing of Conical and Noncircular Gears.

1.5.2 Conventional Gear Finishing

The type and quantity of the microgeometry errors of gears determines their quality which consequently influences the functional performance in terms of noise generation, load-carrying ability, and accuracy in transmission. Errors in microgeometry of gears can be classified in two main types, i.e., form errors and location errors. Profile error and lead error are the main types of form errors, whereas pitch error and runout are types of location errors. In addition, the surface finish quality of the teeth also determines the tribology behavior and wears characteristics. High-surface roughness and the presence of nicks, burrs, cracks, and other surface defects may lead to early failure of gears.

Fundamentally, most conventional gear manufacturing processes is limited to shaping of the required teeth geometry and is unable to impart a good-quality surface finish. Precision applications necessitate high-quality gears for smooth and low-noise operation. This entails inclusion of finishing techniques to obtain the required gear tooth surface finish with acceptable accuracy of the microgeometry. Grinding, shaving, honing, and lapping are material removal-based finishing techniques, whereas rolling is a form-finishing technique. All these techniques are fundamentally used to obtain an adequate surface finish while shaping the teeth to the required geometry while also removing burrs and nicks. Chapter 5, Conventional and Advanced Finishing of Gears, introduces the working principles, mechanisms, and salient features of all the important conventional gear-finishing processes.

1.5.3 Advances in Gear Manufacturing and Finishing

Inherent limitations of conventional processes of gear manufacturing, namely high manufacturing costs, limited quality, and the inability to deal with a wide variety of shapes, sizes, and materials were the main factors behind the development of advanced/modern gear-manufacturing processes. Most conventional manufacturing processes alone cannot produce gears of acceptable quality and therefore requires a finishing technique to further improve the quality. Besides quality aspects, processing of advanced gear tooth geometries and materials are also major concerns where performance of conventional processes is restricted. Conventional gear manufacturing and finishing operations implies extended process chains which are costly and may be less environmentally friendly that desirable.

Advanced processes of gear manufacturing comprise modern methods/processes and advancements to some conventional processes to deal with the abovementioned challenges. Laser beam machining, abrasive water jet machining, spark-erosion machining, metal injection molding, additive layer manufacturing, LIGA etc. are just some of the modern processes used to manufacture high-quality gears from a wide range of materials and of various types and sizes. Their perceived ability to manufacture net-shaped gears can eliminate the need of further finishing, which thereby shortens the process chain, minimizes the production cost, and ensures environmental sustainability. Advancements in conventional methods increase their efficiency to manufacture gears with tight tolerances and improved surface finish. In addition, the concept of environmentally benign green or sustainable manufacturing (i.e., dry hobbing and minimum quantity lubrication-assisted machining etc.) has been introduced in gear manufacturing to produce gears of good quality while maintaining energy, resource, and economic efficiency at the same time. These advanced processes as applied to gear manufacturing are introduced and discussed in Chapter 4, Advances in Gear Manufacturing.

Gear finishing implies removing burrs and nicks, refining teeth surfaces, and achieving geometric tolerance. Manufacturing competitiveness and certain inherent limitations of conventional finishing techniques are the main drivers for the development and use of advanced/modern gear finishing processes. Electrochemical honing (a state-of-art method), abrasive flow finishing, water jet deburring, electrolytic deburring, Thermal deburring, brush deburring, vibratory surface finishing, and black oxide finishing are some of the significant advanced processes for gear finishing. These are presented and discussed in Chapter 5, Conventional and Advanced Finishing of Gears.

The processes such as surface hardening, peening, and coating are used to enhance the surface properties of gears for their improved performance in aspects such as surface hardness, tribology, and residual stress state with the ultimate aim to improve operation in terms of service life and dynamic performance (noise) etc. Chapter 6, Surface Property Enhancement of Gears, presents the aim, methodology, and advantages of various advanced surface modification processes/methods currently employed in gear engineering.

The assessment of the geometric accuracy is an important aspect of the gear manufacturing process. The methods and corresponding instruments employed to measure macro and microgeometry parameters of gears in order to assess their accuracy are presented in Chapter 7, Measurement of Gear Accuracy.

REFERENCES

[1] D. Dudley, The evolution of the gear art, first ed., American Gears Manufacturers Association, Washington D.C, 1969.

[2] The New Encyclopaedia Britannica, 15th ed., Encyclopaedia Britannica Inc., 1992.

[3] D.S. Price, Gears from the Greeks: The Antikythera Mechanism, A Calendar Computer from Ca. 80 B. C., Transactions of the American Philosophical Society, vol. 64 (7), American Philosophical Society, Philadelphia, 1974.

[4] P.L. Fraenkel, Water lifting devices, FAO irrigation and Drainage Paper 43, Corporate Document Repository, Food and Agriculture Organization of the United Nations, Rome, 1986. <http://www.fao.org/docrep/010/ah810e/AH810E08.htm#Fig. 23>.

[5] G.M. Maitra, Handbook of gear design, second ed., Tata McGraw Hill Publishing Company Ltd, New Delhi, 1994.

[6] D.P. Townsend, Dudley's gear handbook, second ed., Tata McGraw Hill Publishing Company Ltd., New Delhi, 2011.

[7] J.J. Coy, D.P. Townsend, E.V. Zaretsky, Gearing, NASA REFERENCE Publication 1152, AVSCOM Technical Report 84-C-15, 1985.

[8] E. Oberg, F. Jones, H. Horton, H. Ryffel, C. McCauley, Machinery's Handbook, Thirtieth ed., Industrial Press Inc., Connecticut, 2016.

[9] J.F. Jones, Gears, in: J.G. Bralla, (Ed.), Design for manufacturability handbook, Tata McGraw-Hill Companies Inc, New York, 2004, pp. 4.253–4.286.

[10] J.R. Davis, Gear materials, properties, and manufacture, first ed., ASM International, Ohio, 2005.

[11] B. Bhushan, Nanotribology and nanomechanics: An introduction, second ed., Springer-Verlag, Heidelberg, 2008.

Chapter 2

Conventional Manufacturing of Cylindrical Gears

The conventional processes for producing gear teeth are grouped into three main classes: subtractive or material removal processes, forming processes, and additive processes [1-3]. Each will be introduced and discussed in some detail.

2.1 SUBTRACTIVE OR MATERIAL REMOVAL PROCESSES

This is one of the most important classes of gear manufacturing techniques for cutting teeth of the required geometry in prepared gear blanks. These are machining processes that employ a cutting tool which is shaped according to the shape and size of the gear teeth to form or generate them in the gear blanks.

Form-cutting and generative processes are two important material removal type processes used to manufacture cylindrical gears.

2.1.1 Form Cutting

In **form cutting**, the geometry (form) of the tool cutting edge replicates the gear-tooth geometry and therefore shapes similar teeth profiles into the gear blanks. These are also referred to as **copying** or **profiling** methods. Milling, broaching, shear cutting, and teeth cutting on shaper are the four main form-cutting processes for cylindrical gears. They are briefly described in the following sections.

2.1.1.1 Gear Milling

Gear milling is an economical and flexible process for cutting a variety of cylindrical and other gear types such as spur, helical and bevel gears, racks, splines, and ratchets. In gear milling, circular, disc-type cutters and end-mill cutters are used to cut gear teeth. The shape of the milling cutter conforms to the gear-tooth space. Each tooth is cut individually; after completion of a tooth, the cutter is returned to its starting position, the blank is indexed for

Advanced Gear Manufacturing and Finishing. DOI: http://dx.doi.org/10.1016/B978-0-12-804460-5.00002-X

the second tooth, and the cycle is repeated. This process is recommended for production of small volumes of low-precision gears. Gears having different modules and number of teeth need separate milling cutters. Milling cutters are less costly than hobs and other types of cutters. End-milling cutters are used for cutting the teeth of large gears of high modules. Cylindrical gears made by milling find applications in low-speed machinery and where the microgeometry deviations of the gears are not of major concern.

Gear milling by disc cutter and end-mill cutter are shown in Fig. 2.1A and B. The disc cutters are used for the form cutting of big spur gears of large pitch, while end-mill cutters are for pinions of large pitch and for double helical gears.

2.1.1.2 Gear Broaching

Cylindrical gears machined by **broaching** possess high geometric accuracy with an excellent surface finish. This machining process removes metal from the gear's blank surface by a forward (push) or rearward (pull) displacement of a multiple-toothed tool known as a **broach** that replicates the geometry of the gear tooth to be cut. Although this process is suitable for high-volume production of both external and internal-type spur and helical gears, it is mostly employed to broach internal gears, racks, splines, and sector gears. Due to the high cost of broaches and requirement of separate broach for each

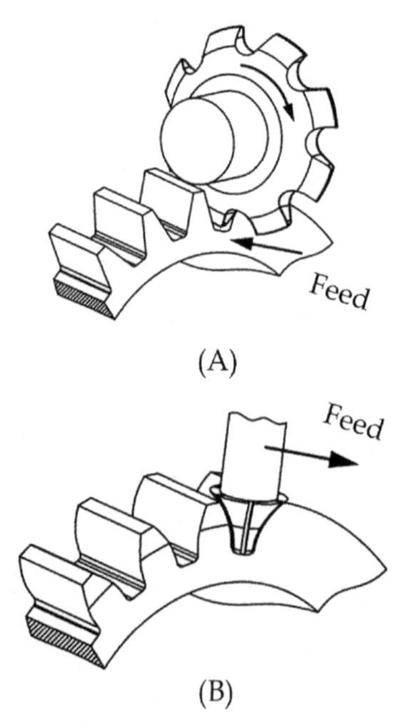

FIGURE 2.1 Gear milling: (A) milling by disc cutter; (B) milling by end-mill cutter.

FIGURE 2.2 Broaching of an internal spur gear.

gear size, broaching is suitable mainly for high-quantity production. A broach tool performs cutting very similar to a single-form tool. The total number of teeth in each section of the broach is equal to the number of tooth spaces to be cut for a particular gear.

Broaching of internal gears requires proper spacing of teeth which may either be equal or unequal, and symmetrical or asymmetrical tooth forms. To withstand the broaching pressure, the tooth forms must be uniform in the direction of the broach axis, and their surface must be strong enough. Most internal broaching is carried out with pull broaches. Single pass of a broach is enough to cut the small internal gears, while large internal gears are usually broached by using a surface type broach that cuts several teeth with a single pass. Broaching has been used to cut gears as large as 1.5 m in diameter. The helical internal gears of automotive automatic transmission are, for example, typically made using a broaching process. Fig. 2.2 illustrates the broaching of an internal gear.

The **pot broaching** method is used for form-cutting external gears. This method consists of a broach tool with internal teeth that are held in a pot that is passed over a blank, producing external cylindrical gears and splines. External gears can also be broached by an inverse type process where the gear blank is displaced (pushed) through a stationary pot broach.

2.1.1.3 Gear Cutting on a Shaper

A **shaper** is machine tool utilizing linear motion for cutting. It is used primarily in gear cutting to manufacture lower quality gears of simple profile, such as spur gears, splines, and clutch teeth. Large quantities of gears may be economically cut with a cutting tool with a cutting edge that corresponds to the shape of the tooth space. The tool reciprocates parallel to the center axis of the blank and cuts one tooth space at a time. Successive teeth are cut by rotating the gear blank through an angle corresponding to the pitch of the teeth until all the tooth spaces have been cut. Fig. 2.3 illustrates the principle of cutting gear teeth by means of a shaper.

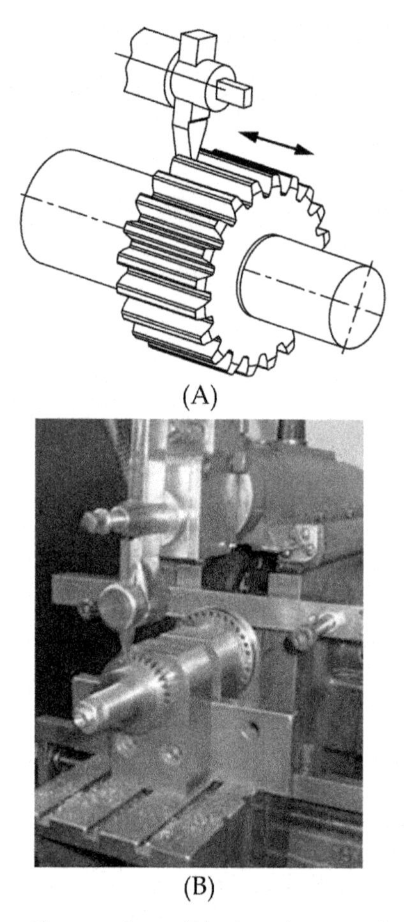

FIGURE 2.3 Gear teeth cutting on a shaper. (A) schematic presentation (B) actual photograph.

2.1.1.4 Shear Cutting

Shear cutting is an important process for machining internal and external spur gears. This process is similar to broaching due to the simultaneous cutting of all the teeth in a linear action between gear blank and cutter. The cutting tool (cutter) resembles a set of form tools held in a special holder. This method is used for high-volume production of coarse pitch gears from 2 to 12 DP. Helical gear cutting is not feasible by this process and it is largely used for external spur gear manufacture [4].

2.1.2 Generative Processes

The **generative processes** are characterized by automatic indexing and the ability of a single cutter to be used to cut gears with any number of teeth for a given combination of module and pressure angle. These processes are known for high productivity while being economical. Gear hobbing, gear shaping, and gear planning are the three major classes of generative type processes for machining cylindrical gears; they are discussed in the following sections.

2.1.2.1 Gear Hobbing

Gear Hobbing is the process of generating gear teeth by means of a rotating cutter referred to as a 'hob'. A hob resembles a worm gear; it has a number of flutes (also referred to as a gash) around its periphery, parallel to the axis, to form cutting edges. The hob is rotated and fed against the rotating gear blank to generate the teeth. Hobbing can be used to produce spur, helical, and worm gears, as well as splines in almost any material (ferrous and nonferrous metals and plastics), but not bevel or internal gears. The versatility and simplicity of hobbing makes it an economical method of cutting gears although conventional hobbing cannot achieve high accuracies (close tolerances) and needs subsequent finishing operations if high accuracy is required. Fig. 2.4 depicts the working principle of the gear hobbing process.

Depending on the direction of feed of the hob, gear hobbing may be classified as either axial hobbing, radial hobbing, or tangential hobbing [5].

2.1.2.1.1 Axial Hobbing

In axial hobbing, firstly, the gear blank is brought toward the hob to get the desired tooth depth then the rotating hob is fed along the face of the gear blank parallel to its axis. This is used to cut spur and helical gears.

2.1.2.1.2 Radial Hobbing

In this type of hobbing, the hob and gear blanks are set with their axis normal to each other. The rotating hob is fed against the gear blank in radial direction or perpendicular to the axis of gear blank. This method is used to cut the worm wheels.

2.1.2.1.3 Tangential Hobbing

It is also used for cutting teeth on worm wheel. In this case, the hob is held with its axis horizontal but at right angle to the axis of the blank. First, the hob is set at full depth of the tooth and then fed forward axially tangential to the face of the gear blank to cut the teeth.

2.1.2.2 Gear Shaping

Gear shaping cuts gear teeth with a gear-shaped cutter mounted in a spindle with its main axis parallel to the axis of the gear blank. The cutter reciprocates axially across the gear blank to cut the teeth, while the blank rotates in

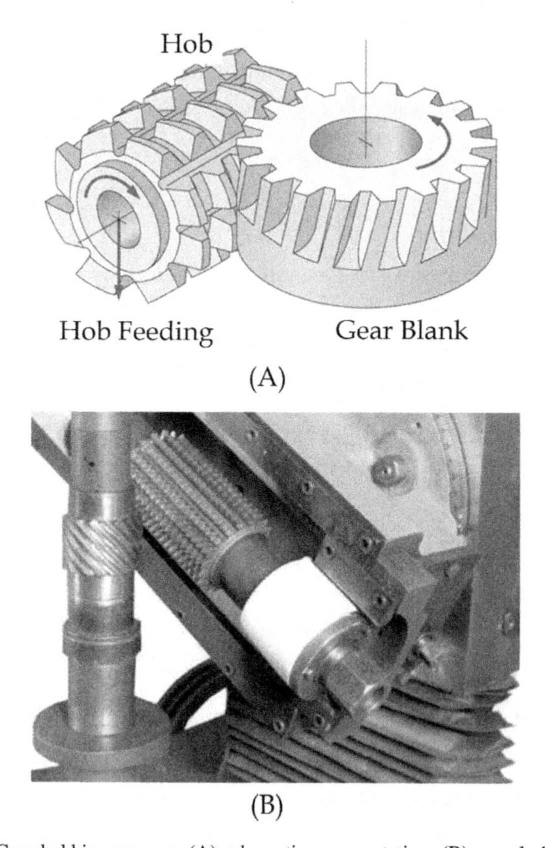

FIGURE 2.4 Gear hobbing process: (A) schematic representation; (B) actual photograph.

mesh with the cutter. The teeth of the gear are generated by the axial stroke of the cutter in successive cuts. The tapered cutter ensures clearance on the sides of the cutting teeth. Cutting occurs on the downward stroke, while during upward stroke, the cutter and the work are moved apart to prevent them rubbing against each other. The principle of gear cutting by shaping is depicted in the Fig. 2.5.

Gear shaping can be used to cut spur gears, helical gears, worm gears, ratchet wheels, elliptical gears, and racks. The biggest advantage of shaping is that it can be used to produce both external and internal gears.

2.1.2.3 Gear Planing

Gear planing is one of the oldest processes for manufacturing spur and helical gears by using a rack-type cutter. The teeth are generated by a reciprocating planing action of the cutter against the rotating gear blank. Gear-planning machines are of two types; one based on the Sunderland

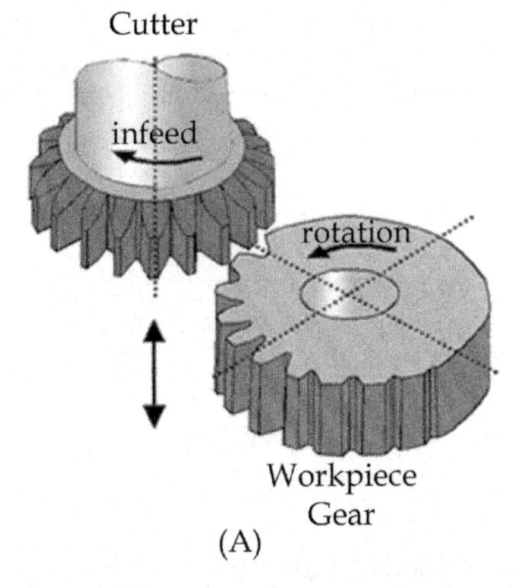

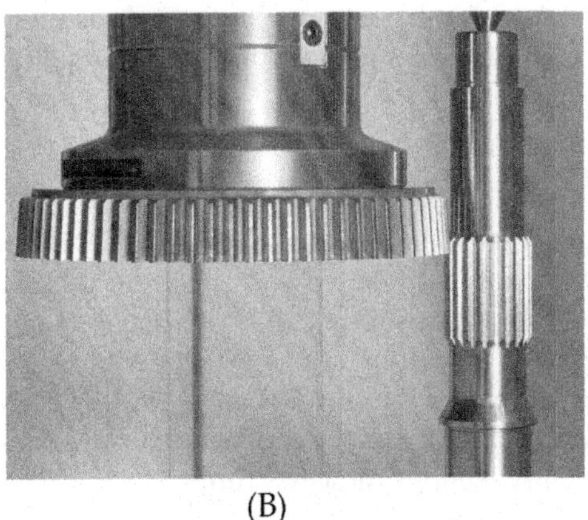

(B)

FIGURE 2.5 Principle of gear shaping: (A) pictorial representation; (B) actual photograph.

process and other on the Maag process, both of which are named after their inventors. They are identical in principle but differ in machine configuration and detail [2,5].

- In the Sunderland process, the gear blank is mounted with its axis in the horizontal plane, while the cutter reciprocates parallel to the axis of the gear blank. The cutter is gradually fed into the gear blank to the required depth. The gear blank rotates slowly while the cutter rack is simultaneously displaced at the same linear speed as the gear circumferential speed. This relative motion brings a new region of the blank and cutter rack into contact, causing the cutter teeth to cut wheel teeth of the correct geometry in the gear blank. A full revolution of the gear blank will therefore require an impractically long cutter rack. To keep the cutter rack to practical lengths, it is disengaged from the blank after the blank has rotated one or two pitch distances. It is retracted to an appropriate position and reengaged with the blank and the process is restarted. This implies that a relatively short cutter rack may be used with the added benefit that all the teeth are basically cut with the same cutter teeth which benefits uniformity.

Fig. 2.6 presents the generation of spur gear teeth by Sunderland process.

- In the Maag process, the gear blank is mounted on the machine table with its axis in a vertical position. The cutter head, incorporating a rack-type cutter, slides vertically in the sides provided at the front of the machine. The cutter can be set at any angle in a vertical plane, and can also be made to reciprocate in any direction. The photograph of Maag gear planer is shown in Fig. 2.7.

This method is less accurate than shaping and hobbing due to the introduction of errors in the tooth geometry by periodical repositioning of the rack and the gear blank for completion of the entire circumference [6].

2.2 FORMING PROCESSES

In all tooth-forming operations, the teeth on the gear are formed simultaneously by a mold or die with the appropriate teeth geometry. The quality of the die or mold is mainly responsible for the accuracy of the teeth. Most of the forming processes are only suitable for high production quantities due to expensive tooling. Manufacturing gears by means of forming has several advantages over material removal processes, for example, significantly shorter process times, lower material loss, and subsequently no chip disposal, and increased strength and higher surface finish in some instances. A number of forming methods such as stamping, fine blanking, extrusion, cold drawing, rolling, and forging are used to make gears by forming the teeth in blanks using dies of the required form

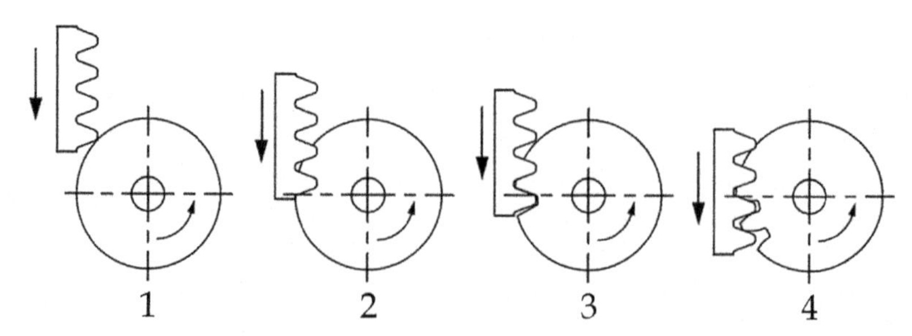

FIGURE 2.6 Generation of spur gear by Sunderland method of gear planing.

FIGURE 2.7 Maag gear planer for cutting spur and helical gears.

and shape. The various forming techniques are discussed in some detail in the following sections.

2.2.1 Stamping and Fine Blanking

Stamping is an economical method for high-volume production (up to 400 gears per minute) of lower quality, lightweight sheet-metal gears of low thickness (0.25–3 mm), which can be employed in low to medium-duty applications such as toys, simple mechanisms, washers, electrical appliances, and water meters. Stamping is a process where sheet or stock metal is cut (sheared) and/or shaped between a tool and a die containing the appropriate geometry. Thin-gauge gears may therefore be relatively easily and cost-effectively produced. A poststamping shaving operation may be required to refine the teeth profiles of the stamped gear. With dies of higher precision, the quality of stamped gears can be improved such that they can be used in timer mechanisms, and other precision applications. Low and medium-carbon steels, brass alloys, and aluminum alloys are the materials of choice for gear stamping. It is possible to attain the American

FIGURE 2.8 Spur gears produced by the stamping process.

Gear Manufacturers Association (AGMA) accuracy numbers as high as 9 through this process [3]. Typical spur gears produced by stamping are shown in Fig. 2.8.

Fine blanking (also known as fine-edge blanking) is a special type of stamping process that produces parts of enhanced accuracy, flatness, and edge characteristics when compared with conventional stamping. Fine blanking makes use of additional supporting tools (pads) and impingement technology to minimize plastic deformation during the shearing process. Fine-blanked gears are used in automotive parts, appliances, office, and medical equipment [1].

2.2.2 Extrusion and Cold Drawing

Extrusion and cold drawing involves forming the required teeth geometry on cylindrical rods referred to as pinion rods. A cylindrical bar of the required material is drawn (pulled) or extruded (pushed) through a series of several dies, the last of which has the final shape of the desired tooth geometry. The cylindrical bar is progressively deformed into the shape of the die to produce the desired tooth form. Since the material displacement takes place by pressure, the outer surface of the gear is relatively smooth and work hardened.

Individual gears are then cut from the extruded pinion rods and may be further finished. Materials such as high-carbon steels, brass, bronze, aluminum, and stainless steel which have good drawing properties can be used to form gears by extrusion [1]. The gears made by these processes possess good surface finish with clean edges. Only straight spur gears can be manufactured by these processes and are majorly used in watches, clocks, appliances, typewriters, small motors, cameras, and simple mechanisms. Brass pinions made by extrusion are shown in Fig. 2.9.

2.2.3 Gear Rolling

In **gear rolling**, a cylindrical bar gear blank is passed through a set of rolls that are shaped appropriately to the gear to be rolled and therefore impart the desired

FIGURE 2.9 Extruded brass pinions.

geometry to the blank. Spur gears, helical gears, worms, and splines are roll formed from a wide range of materials, including carbon steels, alloy steels, and nonleaded brasses. Rolling imparts good surface finish and refines the microstructure of the gear teeth. This process is limited to spur gears that have more than 18 teeth and a pressure angle of not less than 20°. Typically, for successful roll forming of gears, the blanks should not be harder than 28 HRC [1,2].

Cold roll forming (i.e., rolling) of gears may be done in one of two ways, namely flat rolling and round rolling [7–9]. **Flat rolling** functions with two flat-rolling tools moving in opposite directions that mesh with the rolling gear blank symmetrically to the rotation (Fig. 2.10). It is centered between the tips on both ends and can rotate freely. The upper and lower rolling rods have translatory and synchronous motion in relation to one another; they encounter the rolling blank simultaneously and they set it into rotation by means of friction and form closure. Teeth are progressively formed by workpiece deformation with a controlled reduction of the tool relative separation. In this process, the material is displaced at the contact points and flows into the unfilled regions of the tool profile. The correct gear contour including involute flank profiles are formed by the kinematics of the process and an appropriate tool shape [7].

Round rolling of gears is an efficient incremental cold massive-forming process which forms a complete gear into a material. The bored gear blank (rotationally symmetric) coupled with small shaft is clamped in a fixture (in axial direction) which allows its rotational and axial movement. A set of rolling dies/tools (two or three gears of the same profile and geometry) is used to form the gear teeth by compressive loading (Fig. 2.11). The forming

FIGURE 2.10 Flat rolling of a gear [7].

FIGURE 2.11 Round rolling of a gear [7].

process is generally divided into three phases, viz., the first phase in which a tool is punched into the predetermined preforming diameter of the gear; the second is the penetration and reversal of rotation at synchronous feed and rotation speed to roll the exact number of teeth and desired tooth root diameter; and lastly, the calibration of the full-formed gearing profile by optimizing the surface and geometry through some additional form cycles which is followed finally by a release of the finished gear to restart the same process on new gear blank [7–9]. Controlling the material flow and the size of the rolling force remain significant challenges that must be overcome by proper tool design and multistage rolling operations. These gears display some useful properties. Because of the material flow during the process, these gears have a contour following texture orientation. This characteristic leads to improved surface mechanical properties. This process can generate a high-quality surface finish up to 0.4 μm and geometric accuracy up to German standard Deutsches Institut für Normung (DIN) 7 [8].

2.2.4 Gear Forging

Forging has long been used in the manufacture of cylindrical and other gears. In this method, a cylindrical billet of the required material is heated and then forged into a die cavity that has the shape of a finished gear. After being forged, the gear is allowed to cool in air. High-quality gears with an excellent surface finish and high-fatigue strength (due to advantageous texture, and grain flow pattern in the teeth) can be produced by forging with extra care. Forged gears are extensively used in differential gearboxes, agriculture equipment, material handling industries, mining machines, and marine transmissions. Spur gears can be made by forging, but the die life is usually limited. This process is most suitable for bevel and face gears.

2.3 ADDITIVE PROCESSES

As the name implies, in the additive class of gear manufacturing processes, the gear material (in powder or liquid form) is combined or added to produce gears with the help of dies and molds of the required gear geometry and shape. Gear casting, powder metallurgy, and injection molding are the important processes that belong to this class of gear manufacturing. These processes typically include the raw material preparation followed by the actual production of the gear to produce gears of various shapes and sizes from metallic and nonmetallic materials and plastics. A brief description of the important additive gear manufacturing processes, with their unique aspects, is given in the following section.

2.3.1 Gear Casting

Casting is an important process for preparing gear blanks in gear manufacturing. Besides blank production, casting can also be used to produce gears that have teeth of various sizes and forms. Sand casting, die-casting, and investment casting are important methods of casting gears.

- Die-casting is the most extensively used casting process for high-volume production of the small gears used in small machine tools, instruments, cameras, toys, appliances, and electronic devices. This process involves injecting molten metal into a die cavity under pressure. It is then allowed to solidify and removed from the die [2]. Die-casting is followed by a trimming operation for final shaping and sizing of the gear. The two most significant types of die-casting are hot-chamber and cold-chamber casting. The former is limited to processing iron or steel as gear materials. Based upon the application requirements, gears can be die cast from zinc, brass, aluminum and magnesium. (Refer to Chapter 1: Introduction to Gear Engineering for details on materials for the die-casting of gears.) This process is suited to high-volume production of low-cost gears that have limited accuracy.

- Sand casting produces low-quality, large-size gears from cast iron, cast steel, and bronze. They are used for farm machinery, large mills, hand-operated devices, small appliances, lifting mechanisms, and hand-operated cranes. Figs. 2.12 and 2.13 show some examples of sand-cast cylindrical gears.
- Investment casting is another important method, used to produce small-to-medium-sized gears of spur, helical, and even spiral bevel shapes. Its scope is usually limited to the production of gears made from hard materials which cannot be machined, such as tool steel, nitriding steel, and

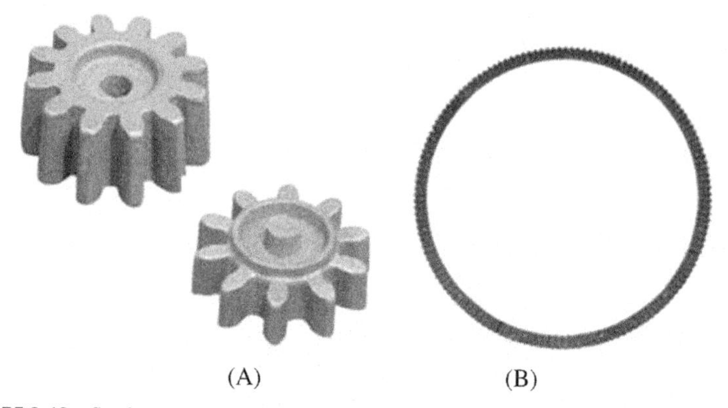

FIGURE 2.12 Sand-cast (A) gears; (B) gear rings to use in a cement mixer.

FIGURE 2.13 Sand-cast steel ring gear to use in mining machinery.

monel. Permanent mold casting is suitable for the production of gears of simple configuration and limited accuracy for use typically in the agricultural industries.

2.3.2 Powder Metallurgy

Powder Metallurgy is a process used to make gears from metallic powders of bronze, iron, stainless steel, and alloy steels. These gears are used in electric motors, automobiles, pumps, power tools, and small appliances.

The process usually consists of three steps. During the first step, appropriately fine powder of the selected workpiece material is sourced and may be blended with other powders or binders. This is then followed with a compaction step, usually at room temperature under high pressure, in a die cavity which has the shape of the desired gear. The last step usually then includes the removal of the compacted blank (referred to as the green state) before it is sintered in a furnace. Depending on the required accuracy, the gears may be subjected to a recompaction or coining operation for final finishing. Cylindrical, spur, helical, and other kinds of gears, such as bevel and face gears, can be made by this process.

Powder metallurgy is economical for mass production of gears of sizes ranging from a few millimeters to over 300 mm in diameter. It can produce gear-tooth accuracies of up to AGMA 8 [2]. In view of limitations in press capacities and die dimensions, this process cannot be used to produce herringbone gears or helical gears with a helix angle of more than 35° [1]. Fig. 2.14 presents various gear shapes made using the powder metallurgy technique.

FIGURE 2.14 Powder metallurgy gears.

2.3.3 Injection Molding of Plastic Gears

Injection molding is the most extensively used process to manufacture plastic gears. It has the capability to produce good-quality, small, nonmetallic gears of various geometries for light-load and noiseless transmissions. Nylon, acetal, polycarbonate, and polyester types of thermoplastic materials are commonly processed by injection molding to make gears. Various additives such as polytetrafluoroethylene (PTFE), silicon, molybdenum disulfide, and graphite are also added to improve strength, inherent lubricity, heat stability, impact and wear resistance, and other properties of the thermoplastic materials [1–3]. Injection-molded parts, including gears, are manufactured by melting a specific, or combination of, powdered or pelletized polymers before injection into a die containing the desired gear geometry of the gear. The full mechanical properties of the polymer and functionality of the gear is then realized by subsequent cooling, solidification, and ejection from the die and final trimming, if required.

Fig. 2.15 and Fig. 2.16 depict an injection-molding die and molded plastic gears, respectively. The processing time depends on the type of plastic and the gear geometry. After solidification, the gears are ejected from the die. This process has high production rates and usually does not necessitate any finishing operation. In general, injection molding can produce plastic gears of AGMA 7 quality. Cameras, printers, toys, electronic equipment, projectors, speedometers, and home appliances are important application areas of injection-molded plastic gears. Currently, microinjection molding is also extensively utilized to produce microgears of metal powders as well as plastics.

FIGURE 2.15 Injection-molding die.

FIGURE 2.16 Injection-molded cylindrical plastic gears.

REFERENCES

[1] J.R. Davis, Gear materials, properties, and manufacture, first ed., ASM International, Ohio, 2005.

[2] D.P. Townsend, Dudley's gear handbook, second ed., Tata McGraw Hill Publishing Company Ltd, New Delhi, 2011.

[3] J.G. Bralla, Design for manufacturability handbook, second ed., The McGraw Hill Companies Inc., New York, 1998.

[4] S.P. Radzevich, Gear cutting tools: Fundamentals of design and computation, first ed., CRC Press, Florida, 2010.

[5] Production Technology, (HMT Bangalore), Tata McGraw Hill Publishing Company Ltd., NewDelhi, 1980.

[6] H.J. Watson, Modern gear production, first ed., Pergamon Press, New York, 1970.

[7] R. Neugebauer, M. Putzl, U. Hellfritzsch, Improved process design and quality for gear manufacturing with flat and round rolling, Ann. CIRP 56 (2007) 307–312.

[8] R. Neugebauer, U. Hellfritzsch, M. Lahl, M. Milbrandt, S. Schiller, T. Druwe, Gear rolling process, process machine interactions, lecture notes in production engineering, SPRINGER, Berlin Heidelberg, 2013, pp. 475–490.

[9] K. Gupta, R.F. Laubscher, J.P. Davim, N.K. Jain, Recent developments in sustainable manufacturing of gears: A review, J Clean Prod 112 (4) (2016) 3320–3330.

Chapter 3

Manufacturing of Conical and Noncircular Gears

The selection of a process to manufacture a gear is based on various important aspects such as type and specification of gear, gear material, surface finish, quality required, and cost concerns. Due to their orientation and shape complexity, the conventional manufacturing of conical and noncircular gears implies the use of certain specialized fabrication processes which differ (in principle of operation and mechanism) from that of the manufacturing of cylindrical gears. This chapter provides an overview of conventional manufacturing of conical and non-circular gears.

3.1 MANUFACTURING OF CONICAL GEARS BY MACHINING

Conical gears are manufactured from a gear blank of which the diameter is continuously varying along its axis. This makes the geometry of conical gears more complex and consequently their fabrication challenging. It also therefore restricts their manufacturing by a limited number of processes due to the complexity and the required flexibility.

Conical gears are typically machined on specialized and most ingenious machine tools using specially designed cutting tools. This necessitates a highly skilled and specialized work force. Fortunately, the cutting action of conical gear cutters is similar to those utilized in some of the basic processes of gear manufacturing that includes milling, hobbing, shaping, and broaching. Various machining processes used to manufacture conical gears can be classified into two main categories, namely generative and nongenerative (see Fig. 3.1).

Generative machining processes are of fundamental and generic nature because a variety of conical gears with different specifications can be manufactured by utilizing different combinations of motions given to cutting tool and workpiece blank. At least one member of every conical gear pair should be manufactured by a generative machining process [1,2]. Most often, it happens to be the smaller gear, i.e., pinion. Conical gear teeth cutting by a generative machining process typically involves the following concepts and sequence:

- An *imaginary generating* conical gear whose tooth surfaces are described by a cutting tool which can be adjusted relative to the axis of the

Advanced Gear Manufacturing and Finishing. DOI: http://dx.doi.org/10.1016/B978-0-12-804460-5.00003-1

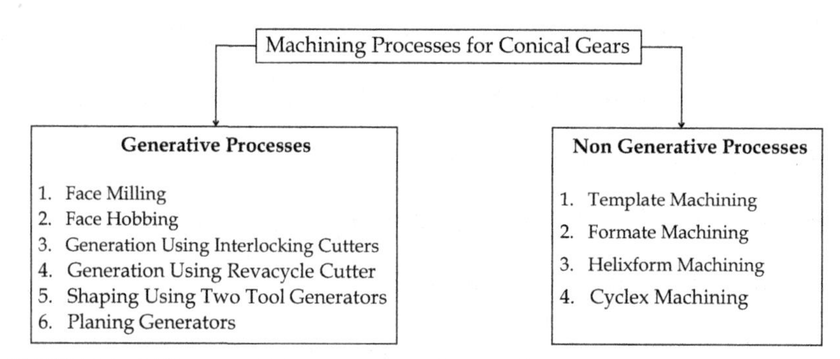

FIGURE 3.1 Different machining processes used to manufacture conical gears.

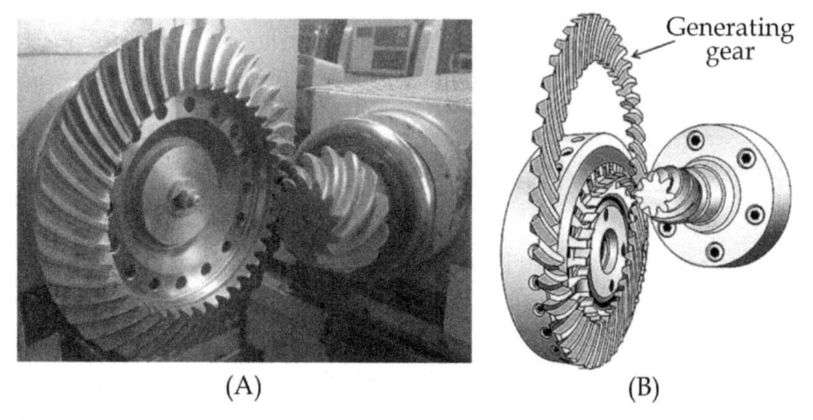

FIGURE 3.2 Concept of imaginary generating gear for a face-mill cutter used in (A) a hypoid gear generating machine and (B) a spiral gear generating machine.

generating gear. Fig. 3.2 illustrates the concept of an *imaginary generating gear* for a face-milling cutter used on the generating machines for a hypoid gear (Fig. 3.2A) and a spiral gear (Fig. 3.2B). Different imaginary generating gears can be used for manufacturing various types of conical gears. These include the following: (1) a member of a complementary crown gear pair, which has a pitch angle of approximately 90°, can be used as the imaginary generating gear for manufacturing both *straight* and *spiral bevel gears*; (2) for manufacturing a *spiral bevel pinion*, a spiral bevel gear or a hypoid gear can be used as the imaginary generating gear; (3) a hypoid gear with a dissimilar pitch and offset angle than its mating gear can be used as the imaginary generating gear for manufacturing a *hypoid pinion*; and (4) for manufacturing a *spiral bevel pinion* or *hypoid pinion*, a gear with a *helicoidal* surface can be used as imaginary generating gear.

- The imaginary generating gear may be a crown wheel that properly meshes with any conical gear with the same diametral pitch and tooth form that of the generating gear.
- The workpiece blank is in the form of a frustum of a cone.
- The cutting tool is mounted on a *cradle* which is a revolving member of the conical gear manufacturing machine and whose axis is coplanar with the axis of the imaginary generating gear.
- Relative positioning of the workpiece blank and imaginary generating gear in such a way that their teeth mesh with each other.
- Either imparting rotatory motion to the cradle and workpiece on their respective axes or making them roll together according to a prescribed motion. In case of bevel gears, this motion is defined by an imaginary nonslip rolling between the pitch surfaces of the generating gear and the workpiece blank.
- Enveloping teeth are cut on the workpiece blank by the teeth of the imaginary generating gear.
- Indexing the workpiece blank so that the required numbers of teeth are cut.

3.1.1 Generative Machining Processes for Conical Gears

Machine tools used in generative machining of conical gears generally employ face-mill cutters, face-hob cutters, or cup-type grinding wheels. The cutter spindle in these machine tools can be swiveled about its axis and can be adjusted radially with respect to the cradle axis. This permits the generation of a variety of conical gears. The ratio of rotation between the spindle of the workpiece blank and cradle is referred to as *ratio of roll*. Generative type machine tools in which the cutter axis is able to tilt are known as *tilting spindle machines*. They maintain a constant ratio of roll during the generative process. Large generative type machines and grinders have non-tilting cutter axis that is affixed in a direction parallel to the cradle axis and are usually adjustable radially. Ratio of roll can be varied during generation process in such machines therefore; these machines are also referred as *modified roll machines*. Machines used for generating straight bevel gears generally have a mechanism to provide crowning on the teeth in which teeth are slightly thicker at the middle than the ends. This avoids concentration of the applied loads at the tooth ends where bevel gear teeth are at their weakest. Face milling and face hobbing are the two commonly used processes for generating teeth on spiral bevel and hypoid gears.

3.1.1.1 Face Milling

The face-milling method makes use of a circular face-mill type cutter with the cradle and workpiece axes rotating in a timed relationship [1]. This flexible method is complemented with a wide variety of cutting cycles. There are basically four face milling methods used to manufacture conical gears. These are (1) single-side face milling, (2) fixed-setting face milling, (3) complete face milling, and (4) single-setting face milling. In each of these processes, the imaginary gear surface is represented by the rotating cutting edges of a face mill cutter.

Single-side face milling cuts only one side of a tooth space in a single operation with a distinct machine setting using a circular face mill cutter which has alternate inside and outside blades, i.e., different machine settings are required to cut both sides of the tooth space.

Fixed-setting face milling cuts the conical gear using two circular face mill cutters. Inside blades of one cutter cuts the convex side of the tooth, whereas outside blades of another cutter cuts the concave side of the tooth. Two machines can be used to cut both sides of a tooth simultaneously to facilitate higher production volumes. Fig. 3.3A depicts generation of hypoid pinion by fixed-setting face milling.

Complete face milling involves generation of conical gears using either a circular face mill or face-hob cutter with alternate inside and outside blades which cut tooth surfaces on both sides of a tooth space simultaneously. This process finishes each gear tooth in one operation. Fig. 3.3B depicts generation of a hypoid pinion by complete face milling.

Single-setting face milling is a variant of complete face milling that is used when the widths of the available cutters are too small to perform the complete face milling in a single operation. Both sides of the tooth space are cut with the same machine setting with the required material removal appropriate for the tooth geometry generation taking place, whilerotating the workpiece gear blank on its axis [2].

Face mill cutters are classified into three types according to the design of the cutting blades, i.e., (1) integral, (2) segmental, and (3) inserted. All three types may be used for both roughing and finishing. The cutters used for

FIGURE 3.3 Generation of a hypoid pinion by (A) fixed-setting face milling and (B) complete face milling.

roughing and complete face milling have either alternate inside and outside blades or end-cutting blades and inside and outside blades arranged alternatively. The cutters used for finishing purposes have either all outside blades or all inside blades or outside and inside blades arranged alternatively. Face milling is used to manufacture *spiral bevel*, *zero bevel*, and *hypoid gears*. Conical gears up to a diameter of approximately 2500 mm are possible to manufacture with current technology, machine tools, and cutters [2].

3.1.1.2 Face Hobbing

Fundamentally, the main difference between face milling and face hobbing is that during face hobbing both the workpiece blank and cutter rotate in strict unison with one another during teeth cutting. This ensures engagement of a group of the cutter blades with a tooth slot as the conical gear is being cut. Face-hobbing cutters have inside and outside blades arranged alternatively. Inserted-blade type cutters are mostly used for face hobbing although integral and segmental cutters can also be used in some situations. As each blade cuts, the workpiece is indexed by one pitch consecutively until all the teeth are completed in a single operation. A typical face-hobbing process is shown in Fig. 3.4.

3.1.1.3 Generation by Interlocking Cutters

This process has two interlocking disk-shaped cutters rotating on their axes that are inclined to the face of the mounting cradle (as shown in Fig. 3.5A). The cutting edges represent a concave cutting surface that removes more material at the tooth ends providing for localized tooth contact. The cutters are mounted on a cradle that rotates them in a synchronous (timed) manner with the spindle of the workpiece gear blank. During the initial rough cutting part of the process, the gear blank is stationary and material removal occurs by movement of the cutter only. This is followed by initiating a fast, in

FIGURE 3.4 Typical conical gear generation by face hobbing. *Note: inserted-blade cutter.

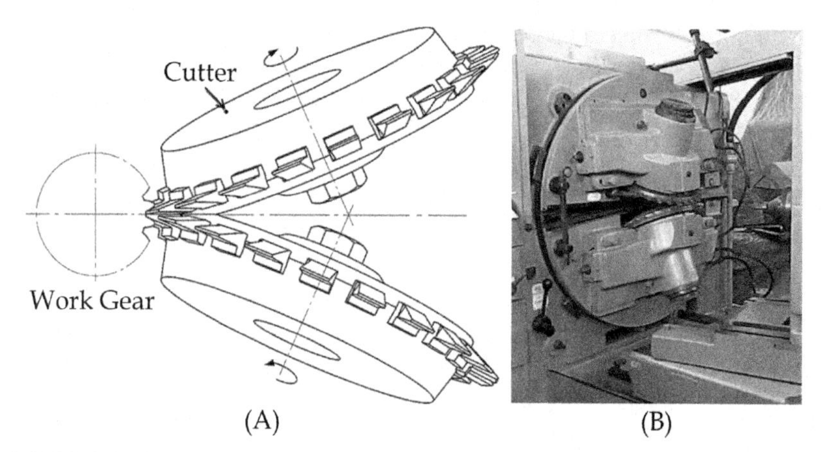

FIGURE 3.5 Concept of conical gear generation by interlocking cutters (A) schematic representation and (B) pictorial view.

unison with the cutter, rolling motion of the gear blank that generates the full depth of the teeth. Thereafter, the blank is retracted and indexed for the next rough cutting cycle. The cycle repeats until all the teeth are cut. The rough cycle may also be completed for all teeth first before the generation cycle commences. Fig. 3.5B presents an example of a typical conical gear manufacturing machine utilizing interlocking cutters.

Interlocking cutters generate the teeth on conical gears or pinions from a solid blank in one operation. These cutters can produce localized crowning also. Unlike other conical gear manufacturing processes, no additional motion is required in this process due to the large diameter of the cutter that facilitates the full generation cycle by simple plunge and roll motions. This method is used to manufacture straight bevel gears.

3.1.1.4 Generation by Revacycle *Cutters*

Conical gear generation by *Revacycle* cutter involves the use of a *circular broach* cutter (Fig. 3.6) in which each successive cutting blade is longer than the preceding blade along its circumference. The convex profiles of the conical gear teeth are generated by the concave edged cutting blades which extend radially outward from the cutter head.

Teeth cutting occurs essentially by the rotating motion of the cutter that induces a broach type motion that is facilitated by the large relative diameter (to gear blank) of the cutter. The progressively longer cutter blades introduce the required feed. A linear relative movement across the face of the gear blank and parallel to its root line is simultaneously introduced. The combined

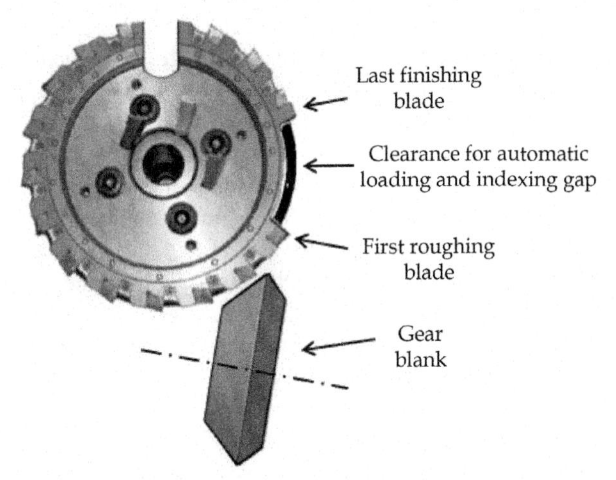

FIGURE 3.6 Generation of conical gears using the *Revacycle* cutter.

effect of the cutter rotation and linear movement across the gear blank face generate the desired tooth shape while facilitating the flat bottom in the tooth space. A single tooth space is generated per revolution of the cutter before indexing to a new tooth space position of the gear blank occurs when the cutter reaches the indexing gap and the process repeats itself.

Revacycle cutters usually consist of three distinct regions that are responsible for *roughing*, *semifinishing*, and *finishing*. The cutting blades may have unique design of side profile relief that facilitates optimum coolant flow. It enhances disposal of chips and improves performance of this process. Separate roughing and finishing operations using different cutters may be required for machining conical gears that require excessive amounts of material to be removed that may not be possible in one operation. Essentially, the roughing cutter then has only roughing blades, whereas the finishing cutter has semi-finishing and finishing blades.

This process is used for high-volume production of straight bevel gears (up to approximately 250 mm pitch diameter) of commercial quality. High production rates make this process the best choice for mass production of straight bevel gears.

3.1.1.5 Shaping by Two-Tool Generators

The two-tool generator machine tool utilizes two reciprocating tools mounted on the cradle that acts as a crown gear. These cutters reciprocate simultaneously in opposite directions cutting the inner sides of two adjacent teeth implying that they cut both faces of one tooth on the gear blank. The correct gear specification and calculation of tooth angle are important for accurate setup of the machine for generation of conical gears having correct angle with respect to their mating gears. Tables provided by the machine tool manufacturers are usually referenced to obtain the relevant setup data. A typical two-tool generator machine tool for conical gear shaping is presented in Fig. 3.7. These machine tools may have the provision to swing the workpiece gear blank spindle at an angle thereby enabling the generation of bevel gears of different cone angles. This spindle is connected to the cradle by a suitable internal change gear set that provide the required rolling motion between the workpiece gear blank and cradle.

Most two-tool generators are used for both rough and finish cutting of straight bevel gear teeth. Separate machines dedicated to rough cutting may be used if production volumes are large, whereas for smaller production volumes both roughing and finishing are usually conducted on the same machine tool. The tooling cost and the production rates of this process are usually lower than those for other straight bevel gear generation techniques

FIGURE 3.7 Photograph of a typical two-tool generator conical gear shaper machine. *Note: The two reciprocating slides on the cradle.

such as interlocking cutters and *Revacycle* cutters. This process is typically used when [1,2]:

- gears sizes are in excess of approximately 250 mm pitch diameter;
- gears with integral hubs or flanges that are problematic to machine by other generation processes; and
- the production quantity is limited and the type or variety of conical gear is problematic to machine by other processes.

3.1.1.6 Planning Generators

Planning generators are capable of producing both straight and curved teeth on bevel gears. Straight, zero, and spiral bevel gears up to approximately 1000 mm in diameter can be successfully cut. It can also be used to generate hypoid gears with the help of special attachments. A straight-edged cutting tool mounted on a reciprocating slide, which is affixed on the face of the cradle, is used to generate the conical gears. Tooth profiles are cut by the rolling movement of the gear blank, whereas the lengthwise shape of the teeth is formed by the motion generated due to the combination of the tool stroke, continuous uniform rotation, and angular oscillation of the gear blank [2]. A typical planning generator machine tool is presented in Fig. 3.8.

One of the major advantages of this machine tool is that it uses relatively simple and inexpensive tools as compared to some of the other techniques. Corrugated tools with 14.5° pressure angles are frequently used.

FIGURE 3.8 A typical example of a conical gear planning generator.

3.1.2 Nongenerative Machining Processes for Conical Gears

Nongenerative machining processes are not based on any particular common fundamental principle or concept rather they use some process-specific concept. Formate, helixform, cyclex, and templatemachining processes belong to this category and are discussed in the following subsections. These processes are generally used to manufacture spiral bevel and hypoid gears particularly when the gear-to-pinion ratio is more than 2.5.

3.1.2.1 Formate Machining

Formate machining can be used for both roughing and single-cycle finishing of spiral bevel and hypoid gears with pitch diameter up to approximately 2500 mm [2]. Both roughing and finishing can be done using one cutter on the same machine but should preferably be performed on different machines for greater accuracy and efficiency. Two types of formatemachining methods are used (1) *formate single-cycle* operation that involves a two-cut roughing and finishing operation in which the workpiece blank is roughed to depth and then finished in a single cycle; or (2) *formate-completing* operation that involves a single-cut roughing and finishing operation.

In the *formate single-cycle* operation, one tooth space is machined in a single revolution of the cutter. Material removal occurs by cutting blades mounted on a circular cutter that resembles a face-milling cutter. Each blade in the cutter is slightly longer and wider than the preceding blade that therefore facilitates effective and continuous material removal. The indexing of the workpiece is ensured by a gap between the first and last cutting blades.

In the *formate-completing* operation, the teeth are cut by two successive actions of the cutter, initially the cutter is plunged to the specified depth (approximately 0.25 mm) before it is then fed to the full tooth depth at higher cutting speed which ensures an acceptable teeth surface quality.

3.1.2.2 Helixform Machining

During helixform machining, the cutter both rotates and translates while the workpiece gear is indexed in an appropriate gap between the cutting blades. This combined motion produces a cutter-blade tip path tangent to the root plane of the gear. A single revolution of the helixform cutter finishes both sides of a tooth space. Helixform cutting is used for nongenerative manufacturing of *spiral bevel*, *hypoid*, and *zero-bevel gears*. This process is advantageous over formate machining in that the gear produced by this process is conjugate to the mating pinion.

3.1.2.3 Template Machining and Cyclex Machining

Template machining is a limited volume production process that is used to manufacture a wide variety of large-sized coarse-pitched bevel gears [2].

A simple single-point cutting tool is guided by a set of templates (one for each side of the tooth space). A good quality finished gear is made in 5–6 cutting operations.

In the cyclex machining, both roughing and finishing cuts are done in a single setting of the gear blank. The finishing blades are placed below the roughing blades and only come into contact with the gear blank when the last roughing blade has passed. The workpiece gear blank is withdrawn after the finishing cut, indexed, and the cycle is repeated until the cutting of all teeth is completed [2]. This method is capable of achieving good accuracy of tooth spacing, a fine finish and high production rates.

Table 3.1 presents a summary of different machining processes used to manufacture various types of conical gears, also highlighting their capabilities and limitations, with the aim to facilitate the selection of an appropriate process.

3.2 MANUFACTURING OF NONCIRCULAR GEARS

Noncircular gears are specially designed gears having unique characteristics and asymmetric shapes, i.e., elliptical, eccentric circular, oval or internal noncircular, etc. usually to fulfill a specific purpose. Applications include producing a unique speed and axis distance profile, stop-and-dwell motions, combined rotation and translation and to maintain constant speed segments in an operating cycle, etc.

These gears present formidable challenges to the gear manufacturing industry due to their complex design and manufacturing difficulties. Recent advances in modeling and simulation software, computer numerical control (CNC) machine tools, and nonconventional manufacturing processes have made their design and manufacturing more feasible and simpler [3].

The concept of manufacturing noncircular gears may be simplified by the effective use of existing theories and methods. An example is the enveloping theory developed by Litvin et al. [4] that puts forward the idea of using the same tools as used in the manufacturing of circular gears to manufacture noncircular gears also. This theory is based on the concept of obtaining the required tooth surface as an envelope of the tool surfaces and conjugate tooth shape by imaginary rolling of the tool centrode over the given gear with appropriate use of the kinematics. Manufacturing of the noncircular gears is also facilitated by the method developed by Riaza et al. [5] which is based on the effective use of Bezier curves to obtain the relationship between the mating noncircular gears.

The manufacturing of the noncircular gears is challenging when compared to circular gears due to the complexity in their geometry and need to fulfill special transmission requirements. Manufacturing processes such as generation by copying and rolling, indexing method, template method, NC gear shaping, and wire electrical discharge machining (WEDM) are generally used to fabricate different types of noncircular gears [6,7].

TABLE 3.1 Summary of Applicability of Different Machining Processes for Conical Gears

Type of conical gear	Machining process	Applicability
Straight bevel gears	(a) Gear shaping using two-tool generators	1. Machining straight bevel gears in a wide range of sizes up to approximately 900 mm (i.e., 35 in) outside diameter 2. Low to medium production volumes (up to 150 pairs of gears) because of low tooling cost. For large production volumes this process becomes prohibitively expensive 3. Machining of gears that have some protruding portion, i.e., front hubs which prohibits use of some other processes
	(b) Generation using interlocking cutters	1. Machining straight bevel gears in a single operation 2. Machining of gears without a front hub and with outside diameter less than approximately 400 mm (i.e., 16 in) 3. Faster than two-tool generator process but more costly 4. For medium-to-high production volumes (up to 1200 pairs of gears)
	(c) Generation using *Revacycle* cutters	1. Fastest production process for straight bevel gears 2. Machining of gears without front hubs and pitch circle diameters up to approximately 250 mm (i.e., 10 in) with 4:1 speed ratio with pinion and face width up to approximately 30 mm 3. Economical mainly for high production volumes due to high tooling cost
Spiral bevel, zero bevel, and hypoid gears		1. Spiral bevel, zero bevel, and hypoid gears are manufactured by using similar machining processes and principles as those used for straight bevel gears. Spiral bevel pinions are however manufactured by generative processes, whereas the gears are manufactured without generating roll specifically when the gear-to-pinion ratio is in excess of 2.5 2. Nongenerative processes such as formate completing, helixform, and cyclex machining are generally used for cutting of spiral bevel and hypoid gears 3. Nongenerated gears are usually less expensive than generated gears 4. The cost difference between the various processes used to manufacture spiral bevel gears up to outside diameter of approximately 800 mm (i.e., 32 in) is usually less significant when compared to the manufacture of straight bevel gears

Until the advent of computer-controlled machine tools, profile copying was a popular method of repeatedly reproducing complex shapes. The *copying process* to manufacture noncircular gears involve tracing along a 2-dimensional sheet metal template of the required shape or the surface of a carefully made prototype with a stylus. The resulting movement is then transferred either electronically or hydraulically to the cutting tool causing it to mimic precisely the motion of the profile of the template or stylus. The required 3-dimensional profile is generated on the gear blank that rotates and/or translates linearly as appropriate. The cutting tool has no influence on the shape produced in a similar manner as for pure generation [6]. Noncircular gear manufacturing using copying is simpler to use and does not require complex tools but has certain inherent limitations. These include (1) a special copy is required for every gear geometry and the manufacturing and installation errors of the copy will be transmitted to the manufactured noncircular gears; (2) gears with internal crowns and those with convex–concave segments cannot be cut; and (3) cutting usually requires many passes.

The *template method* of manufacturing noncircular gears uses the gear planer in the same manner as for cutting conical gears. In this process, cutting of gear tooth profile takes place due to a reciprocating motion of the tool which is similar to a side-cutting shaper tool and controlled by template that guides and reciprocates the tool. The reciprocating tool is mounted on the frame of the gear planer and is guided at one end by a roller acting against the template. The outer end is pivoted at a fixed point. The gear cutting operation is usually completed by the use of three sets of templates (first set for rough cut and two subsequent sets for finishing). In this process, the gear blank is held stationary whereas the tool reciprocates. This process is suitable for cutting large noncircular gears.

Shaping of a noncircular gear is similar to the rack-type cutting process except that the linear-type rack cutter is replaced by a circular cutter where both the cutter and the blank rotate as a pair of gears in addition to the reciprocation of the cutter.

WSEM, also known as wire spark erosion machining, is an advanced thermal type machining process which can also be used to manufacture noncircular gears. In this process, a thin metallic wire acts as a tool to cut the required gear profile. With the help of a computer-aided manufacturing program, the geometry of almost any gear shape is transferred to a tool path that the wire electrode tool then follows to produce the gear. A more detailed process description along with its salient features is presented in Chapter 4.

REFERENCES

[1] D.P. Townsend, Dudley's Gear Handbook: The Design, Manufacture and Applications of Gears, Tata McGraw Hill Education Pvt. Ltd, New Delhi, 2011.

[2] J.R. Davis, Gear Materials, Properties, and Manufacture, ASM International, Novelty, OH, 2005.

[3] M. Vasie, L. Andrei, Analysis of noncircular gears meshing, Mech. Test. Diagn 4 (II) (2012) 70–78.

[4] F. Litvin, A. Fuentes-Aznar, I. Gonzalez-Perez, K. Hayasaka, Noncircular Gears: Design and Generation, Cambridge University Press, Cambridge, UK, 2009.

[5] H.F.Q. Riaza, S.C. Foix, L.J. Nebot, The synthesis of an N-lobe noncircular gear using Bezier and B-spline nonparametric curves in the design of its displacement law, J. Mech. Des 129 (9) (2006) 981–985.

[6] T.F. Waters, Fundamentals of Manufacturing for Engineers, Taylor and Francis, London, UK, 2003.

[7] F. Litvin, A. Fuentes-Aznar, I. Gonzalez-Perez, K. Hayasaka, Noncircular Gears: Design and Generation, Cambridge University Press, Cambridge, UK, 2009.

Chapter 4

Advances in Gear Manufacturing

Traditional gear manufacturing involves the use of typical cutting tools such as cutters, hobs, and other multipoint tools which quickly remove material from the gear blank being machined. It also involves the use of dies to form and/or to add various materials to the gear shapes. The traditional manufacturing processes of gears usually address a specific market need. The scope of traditional manufacturing processes is usually limited to the manufacture of gears for general purpose requirements. Gear manufacturing for precision and ultraprecision applications usually rely on the traditional process for the initial shaping after which a dedicated finishing operation to resize and refine is required. In other words, the traditional manufacturing processes are limited when it comes to net-shaped or near-net-shaped gears. This necessitates subsequent finishing processes to impart the desired quality to gears. These subsequent finishing processes themselves require fabrication, repair, and maintenance of the finishing tool; consumes large amounts of cutting fluid and energy; increases the burden of handling, recycling and disposal of waste and escalates the overall cost.

The advanced processes therefore attempt to address these limitations. They typically involve more modern manufacturing processes and advancements in the conventional processes. The main objectives for developing more advanced manufacturing processes and advancements in the conventional processes are:

- To meet the stringent quality requirements for precision gear applications;
- To cut down the cost of gear production;
- To permit gear designers greater freedom in selection of materials, sizes, and shapes;
- To minimize the consumption of harmful cutting fluids, energy, and other resources;
- To eliminate waste; and
- To ensure energy, resource, and economic efficiency with a lower environmental footprint.

Advancements in conventional manufacturing processes involve the technological advancements in machine tool features and functions, use of improved cutting tool materials, coatings and geometries, and application of

Advanced Gear Manufacturing and Finishing. DOI: http://dx.doi.org/10.1016/B978-0-12-804460-5.00004-3

environment-friendly cutting fluids. Modern or advanced manufacturing processes for gears have been developed using novel concepts, mechanisms, cutting tools, and energy. These processes are aimed at eliminating the necessity of postfinishing by manufacturing net-shaped or near-net shaped gears along with meeting the aforementioned objectives.

Advanced manufacturing processes for gears are classified into four major categories according to the shaping mechanism utilized. They are: (1) subtractive or material removal processes; (2) additive processes; (3) deforming processes; and (4) hybrid processes. Selected advanced manufacturing processes in each of these categories are described in the following sections.

4.1 SUBTRACTIVE OR MATERIAL REMOVAL PROCESSES

4.1.1 Laser Machining

4.1.1.1 Introduction

Albert Einstein, no less, described the theoretical foundations of the *laser* and *maser* in 1917. It would take approximately another 40 years, in 1960, that the first working LASER (light amplification by stimulated emission of radiation) was invented and used in the United States by Theodore H. Maiman (solid ruby laser) [1–3]. Since then, there has been a continuous development in the field of laser technology that facilitates its use in various areas of science, engineering, medicine, food processing, and various other activities.

The ability of high power lasers to cut complex shapes with precision and little material wastage makes them one of the most economical ways to manufacture gears of various types and forms [4]. The laser beam being essentially a massless tool is not subject to wear and tear and offers significant flexibility when combined with automation as an advanced gear manufacturing process. It is more often used for microfabrication of gears made of various materials ranging from lightweight plastics, ceramics, hard, and durable metals and alloys. The exact nature of the interaction and consequently the gear quality is specific to the material and laser processing parameters.

4.1.1.2 Working Principle, Process Mechanism, and Significant Process Parameters

The use of a laser beam for the fabrication of gears is one of the most progressive techniques that slowly replacing certain conventional manufacturing processes. It is most frequently used to manufacture miniature gears. It typically offers an excellent balance between speed, cost, and accuracy. Fig. 4.1 depicts the general principle of gear manufacturing by a typical laser system [5,6]. Essentially it utilizes the high energy density of a laser beam to rapidly ablate (vaporize) a material irrespective of its thermal, physical, and chemical properties. It is a localized process that limits the thermal effects significantly.

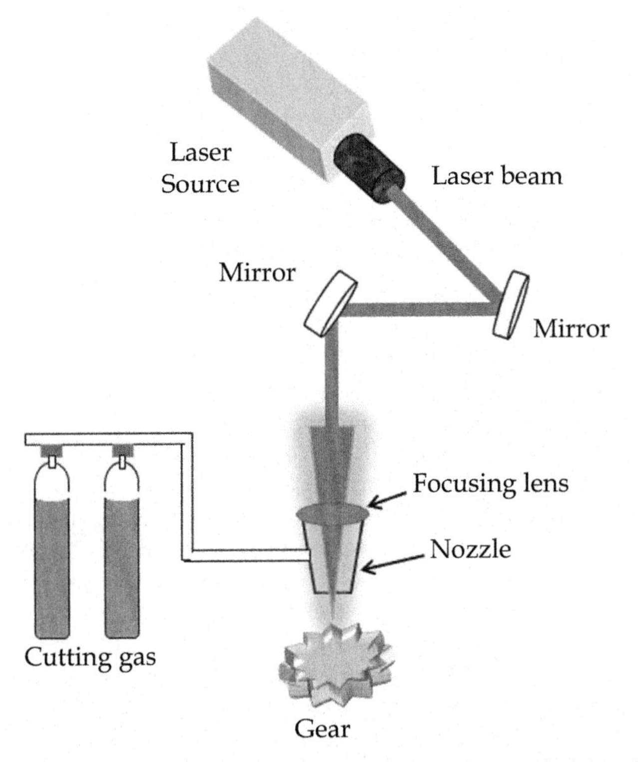

FIGURE 4.1 Schematic representation of a typical laser system for manufacturing of gears.

As shown in Fig. 4.1, the laser beam, generated in the laser source, is directed towards the cutting head or nozzle by a set of mirrors, where it is focused by a lens onto a localized region of the workpiece (gear blank). The intense laser beam quickly heats up the gear material to beyond its melting temperature. The assisting gas, also referred to as the cutting gas, protects and cools the focusing lens while also removing the molten material from the kerf. Typically carbon dioxide lasers, excimer lasers, solid state lasers such as Nd:YAG and Yt:YAG are used for gear cutting.

The intensity of the laser beam (ratio of the laser power to the focussing area) is one of the most important process parameters of laser machining. High intensities are desirable as it leads to increased cutting rates and excellent quality due to the rapid heating and subsequent rapid heat dissipation due to the large temperature gradient. Nozzle geometry and stand-off-distance (SOD) are two other important parameters that may have a significant effect on the cutting ability of a laser. Typical nozzle diameters of 0.8–3 mm and SODs of 0.5–1.5 mm are used. Turbulence and pressure variations may occur in the cutting gas jet for larger values of the SOD. The shape of the focused laser beam is defined by the focal length of the focussing lens. A lens with a shorter focal length usually produces a smaller spot size and a greater laser intensity resulting in higher productivity and improved quality. Typically oxygen is used as cutting gas for gears manufactured from carbon steel, low alloy steel, brass, and other copper alloys, while nitrogen is best suited to machine gears from stainless steel and other high alloy steels.

4.1.1.3 Laser Systems for Gear Manufacturing

A wide range of laser machining systems are available for both micro and macromachining, related research, and development of processes and product. A standard range of the typical laser cutting machines includes large machines that are able to machine both micro and macrogears, mini laser systems for only microgear cutting, and portable desktop CO_2 laser systems with the capacity of few watts, suitable for cutting smaller and especially thin gears in wood and plastics [7]. Fig. 4.2 presents a typical laser cut spur gears machined from 5 mm thick stainless-steel plate.

The thickness of the gear being cut during laser machining may have a significant effect on the quality and is largely responsible for the compromise between quality and accuracy. This implies that typically only smaller gears are laser machined with the average thickness being between 3 and 5 mm. Moreover, other adverse effects associated with laser machining, i.e., heat-affected zones (HAZs) and burr formation may also occur. Short pulse durations (nano, pico, and even femto-second) may be the best options to minimize the thermal effects and thus eliminating the necessity of any postfinishing operations [8,9]. Ultra-short pulse durations imply nearly adiabatic heating of the substrate which allows the substrate surface temperatures to quickly reach the point of vaporization with minimal heating effects on the surrounding areas.

Efforts are continuing as far as laser technology development for gear machining is concerned for producing improved quality and to enable miniaturization [10–12]. These include the fabrication of steel microgears with a module of 0.038 mm and high ratio of speed reduction while motion transmission using an Nd:YAG-IR laser of extremely short wavelength [10]; cutting of ratchet gears of stainless steel deploying a femto-second laser system with high dimensional accuracy and a little recast material (Sandia National

FIGURE 4.2 Close-up view of a laser machined miniature spur gears.

Laboratory, NASA) [11]; and machining of high accuracy polyimide micro-gears by short-pulsed frequency tripled Nd:YAG laser using pulse duration of 500 ps, etc. [12].

In essence, laser machining is a developing technology that facilitates component miniaturization and improved performance applicable to a wide range of materials. This makes laser machining attractive and useful for micromachining, prototyping, and development, as well as for small-to-medium volume manufacturing of gears. The exact nature of the interaction and consequently the gear quality is specific to the material characteristics and laser processing parameters.

4.1.2 Abrasive Water Jet Machining

4.1.2.1 History and Developments

The use of a high-pressure water jet was introduced to the industry largely as a cutting tool in the 1970s (United States). Its initial applications were limited to cleaning and certain destructive applications such as demolition of concrete structures, tunnel boring, and hydraulic mining. From the 1980s onward the use of a water jet as carrier medium for abrasive particles dramatically improved its cutting and drilling capabilities thus enabling machining of harder and/or higher strength materials. Subsequently the abrasive water-jet cutting industry expanded to such an extent that it is now competing on equal footing with established machining processes such as laser beam machining, electric discharge machining (EDM), ultrasonic machining, conventional milling, turning, and other traditional machining processes. Abrasive water jet machining (AWJM) continues its rapid growth as a viable option for gear manufacturing. Its key attributes are: ability to cut gears of almost any material (i.e., flexibility), shape, size, and thickness; ability to machine gears made of composites and other fibrous materials without delamination and burrs; and the ability to produce gears with no thermal distortion and/or damage and excellent surface finish and close tolerances. Short setup times and simple tooling make it a good match for producing prototypes and small-to-medium runs of custom gears [13].

4.1.2.2 Working Principle, Process Mechanism, and Significant Process Parameters

AWJM combines abrasive jet machining with water jet machining (WJM) to create a unique process which overcomes their individual limitations and enhances the machining capabilities of WJM for cutting, drilling, and general cleaning of hard and/or strong materials. In essence it induces material removal by the erosive action of a high-velocity abrasive-laden water jet ejected from a small diameter nozzle onto the target workpiece surface in such a way that the particle velocity is virtually reduced to zero on striking the surface. The heart of the AWJM system is the abrasive water jet (AWJ).

It is a slurry-based cutting jet formed by introducing and entraining small abrasive particles into a high pressure, high-velocity water jet in such a manner that some of the momentum of the water jet is transferred to the abrasives to create a coherent AWJ that exits the AWJM nozzle (see Fig. 4.3). This slurry jet has the ability to machine various hard and/or strong materials such as steel, titanium alloys, honeycomb structures, rocks, stones, ceramics, composites, and concrete up to a thickness of 20 cm at relatively high speed. This is made possible by a combination of the initial particle impact and the subsequent secondary acceleration and impact of a significant fraction of these particles. An AWJM system consists of the following:

- *Pumping unit*: Its aim is to produce a high-pressure water jet that will ultimately transfer its momentum to the abrasive particles. It typically consists of five components namely: (1) *Electric motor:* Typical capacity, approximately 15–56 kW (i.e., 20–75 HP); (2) *Hydraulic pump* driven by the electric motor unit to generate the hydraulic pressure in the range of approximately 15–30 MPa. A hydraulic intensifier pump and/or a crankshaft pump are the two main types typically used. The latter is typically more efficient and therefore produces more net power for faster cutting; (3) An *Intensifier* to generate high-pressure water as high as 40 times more than the hydraulic pressure by using a larger size oil piston than the water piston. The ratio of water pressure to hydraulic pressure is equal to the square of the oil and water piston diameter ratio. Water pressure is easily adjusted by regulating the lower pressure oil; (4) An *Accumulator* is simply a pressure vessel to store high-pressure water to avoid pressure variations

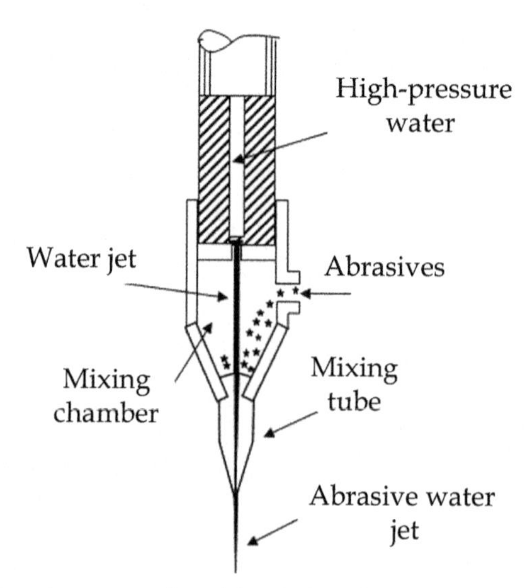

FIGURE 4.3 Schematic representation of functioning of an abrasive water jet (AWJ) nozzle system.

and to produce a smooth and uniform water flow at the output; and (5) *Tubing*: Typically flexible tubing is used if water pressure is less than 24 MPa otherwise rigid tubing is used. A manually or automatically operated on-off valve is installed in the high-pressure line to provide on-off control of the water flow.

- *Water jet unit*: to produce a high-velocity water jet (up to 900 m/s) by passing the pressurized water (even up to 500 MPa) through an appropriate nozzle. This unit is essentially the same as used in the WJM process. Synthetic sapphire, tungsten carbide, and hardened steel are the most commonly used materials for the water jet nozzle.
- *Abrasive feed unit*: It consists of a hopper and flow control system to deliver a precisely controlled stream of abrasive particles to the AWJ nozzle. Typically two particle delivery methods are used: (1) *dry abrasive delivery* which is used for a shorter delivery distance; and (2) *abrasive slurry feed* which introduces the particles over a longer distance. The latter requires more power for a given cut and commercial availability is limited.
- *AWJ nozzle*: to provide efficient mixing of abrasives and water and to form a coherent high-velocity jet of water laden with abrasive particles. The two configurations typically used are: (1) *single-jet side feed* in which the water jet is located centrally and abrasives are added on the periphery of the water jet. It is relatively easy to machine and can be used as a water jet nozzle also. It does however not provide optimal mixing of the water jet with the abrasives and wears out at an increased rate; and (2) *multiple-jet central feed* in which abrasives are introduced centrally with multiple water jets being added on the periphery of the abrasive jet. It provides optimum mixing and typically an increased nozzle life but is costly and more difficult to fabricate. Tungsten carbide and boron carbide are commonly used to manufacture the nozzles. PC-based advanced motion controllers are used to control the AWJ nozzle movement and precise functioning of other systems.
- *Worktable*: The worktable is typically computer numerically controlled (CNC) along the X and Y axes to provide accurate relative motion between the AWJ nozzle and the workpiece to facilitate accurate cutting. Based on the size of the gear to be machined and the size of the gear blank to be used, worktables are available in a wide sizes ranging from small (700 mm × 600 mm) to very large (8000 mm × 4000 mm) [13,14]. Previous generation AWJM systems were often inaccurate and difficult to control but newer systems with modern software-based advanced controllers, flexible attachments, and rugged machine structures offers substantial versatility along with impressive cutting speeds and tolerances [13].
- *Catcher*: to minimize noise by dissipating the energy of the AWJ and to contain the AWJ exiting from the workpiece. A catcher is used for a stationary AWJ nozzle and movable workpiece setup while a settling tank is used when the workpiece is held stationary and AWJ nozzle is moved.

Visual examination of the AWJM process has revealed that material is removed by a cyclic penetration process where work material is removed by erosion through cutting wear at top surface and deformation beneath the surface [15]. The process parameters that have a significant effect on the performance of the AWJM process can be grouped into the following four groups:

- *Hydraulic parameters*: these parameters include water jet pressure, water flow rate (or water jet nozzle diameter). A minimum critical value of water pressure (abrasive velocity or kinetic energy) is required below which no significant machining will occur for each specific gear material. Water flow rate is directly proportional to the square root of the water pressure and the square of the nozzle diameter. The depth of cut increases with an increase in water mass flow rate with a decreasing slope towards a saturation point. The depth of cut varies linearly with water jet nozzle diameter for a given water pressure [15]. Nozzle diameter ranges from 0.075 to 0.635 mm.
- *Abrasive related parameters*: these parameters are basically the type, size, shape, and mass flow rate of the abrasive particles. Abrasives should be harder than the workpiece material. Garnet, silica sand, silicon carbide, alumina, glass beads, and steel grit are the most commonly used abrasives. Garnet is the most effective; silica sand is cheaper but least effective while silicon carbide is expensive but gives effective penetration. An optimum abrasive particle size exists for a particular gear material and nozzle mixing chamber configuration. The range of particle size for optimum depth of cut is wider for brittle materials than that for ductile materials. Commonly used abrasive sizes are 100–150 Mesh number. An optimum value of abrasive mass flow rate exists and any increase in abrasive mass flow rate beyond this value decreases the depth of cut. Generally, abrasive flow rates are 0.002–0.08 kg/s.
- *Cutting parameters*: Traverse or feed rate of the AWJ nozzle, stand-off-distance (SOD), number of passes, angle of impingement, and properties of the gear material are the important cutting parameters.
- *Mixing parameters*: Method of supplying abrasives (i.e., forced or suction), abrasive condition (i.e., dry or slurry) and mixing chamber dimensions.

4.1.2.3 Machining of Gears by Abrasive Water Jet Machining

Advancements in computer technology have greatly benefited the development of WJM and AWJM. Development of PC-based CAD/CAM software and hardware (i.e., controller) has made gear cutting possible with AWJM. It facilitates the creation of gear geometries by CAD software, conversion of said CAD geometries into an executable CAM module, optimization of tool paths, calculation of the feed rate of the AWJ nozzle and decision making as regards to automatic compensation for jet lag and taper [14]. This greatly reduces the operator's input and ensures efficient machining to obtain accurate gears.

Furthermore, micro-AWJM process enables the manufacturing of mesogears and microgears. This requires downsizing the nozzle diameter and abrasives. Smaller nozzles and abrasives presents challenges with nozzle clogging due to the accumulation of wet abrasives and must be addressed to obtain good quality small gears. Typically the rule of thumb is that the maximum abrasive size should be no larger than one-third of the AWJ nozzle inner diameter to prevent clogging due to the bridging of two large abrasives [16].

In order to produce gears for applications requiring close tolerances and a good surface finish the cutting head must be flexible enough to tilt to compensate for the natural taper of the jet. Special tilting head accessories are available that enables near-zero taper to be achieved in gears. Moreover, the optimum parameter combination to prevent "jet-lag" (the difference between the position of the exiting jet and the actual position of impingement) is also an essential requirement in AWJM of gears.

The size of gears that can be manufactured by AWJM ranges from miniature gears to giant mill gears irrespective of its geometry and material. Common gear geometries that can be machined by AWJM include spur, helical, bevel, internal, planetary, and noncircular. Racks, splines, and sprockets can also be manufactured by AWJM at a lower cost than the conventional manufacturing processes. AWJM have been predominantly used for manufacturing of a multivariety of gears made from different difficult-to-machine materials such as composites, polymers, titanium alloys, honeycomb materials, and others. AWJM have been used to successfully manufacture gears as an example from the difficult-to-machine polymer G-10 (plastic resin composite) as used in lapping machines, titanium rack, and pinion gears for commercial jet pilot seats and fine-pitched steel gears having extensive applications in industrial machines [4].

Fig. 4.4 presents some of the gear types machined by AWJM.

4.1.2.4 Advantages of Abrasive Water Jet Machining for Gear Manufacturing

AWJM typically offers the following advantages when applied to gear machining [16]:

- Machining of virtually any type of gear (i.e., external, internal, bevel, noncircular) and almost any thickness;
- Applicable to almost all gear materials irrespective of hardness;
- Generates burr-free edges and good surface quality;
- Tolerances up to ± 0.025 mm or better can be achieved by AWJM;
- Almost no heat generation during cutting hence no HAZ on gear tooth surfaces which may occur in laser machining and spark erosion machining (SEM) of gears;
- No mechanical stresses in gears which often occurs with conventional manufacturing processes;

FIGURE 4.4 Various gear types machined by AWJM.

- Safe operation, no hazardous vapors, and toxic fumes;
- Fast setup and programing;
- Simple fixtures and tooling;
- Cost-effective and productive to machine both small and large gears. For simultaneous machining of identical gears a multinozzle platform can also be used;
- Environmentally-friendly: generates minimum waste with no hazardous waste by-products, and energy and resource efficient.

In a nutshell the technological suitability of AWJM makes this process one of the most versatile, inexpensive, and environmentally-friendly process for gear manufacturing. Efficient AWJM machines and appropriate combination of process parameters can manufacture gears of high quality and accuracy which may eliminate the necessity of postmanufacturing processing. The advancement and refinement of AWJ technology continue to further downsize the gear size and manufacturing cost and to further improve their quality.

4.1.3 Spark Erosion Machining

As previously mentioned, most conventional gear manufacturing processes have inherent limitations and will usually require a subsequent finishing operation to manufacture a gear of appropriate quality. These finishing processes increase energy consumption, material, and cutting fluids and therefore have a negative effect on overall sustainability [17]. Potential single-stage manufacturing of gears by *SEM* seeks to overcome some of the drawbacks of traditional manufacturing processes by improving the gear quality, and attaining improved energy and resource efficiency. SEM and its variants such as wire spark erosion machining (WSEM), micro-SEM and micro-WSEM have been acknowledged as potentially significant substitutes for conventional processes because of their excellent repeatability, geometrical accuracy, surface integrity, quality, reduced setup time, being able to run unattended for long durations, ease of cutting complex shapes and geometries, elimination of mechanical stresses during machining, ability to cut any electrically conductive material irrespective of its hardness, toughness or melting point, and mostly avoiding the need of subsequent finishing operations [18–21].

4.1.3.1 Introduction and History

The phenomenon of material erosion by an electric spark was first noticed by Joseph Priestly in 1878, but this concept could not be used for machining until the 1940s when the Lazarenko brothers in Russia designed a circuit for controlling the electric sparks for machining purposes. Since then *SEM* also referred to as *EDM* has been one of the most widely used thermal type advanced machining processes. It removes the workpiece material by a controlled erosion due to melting and vaporization caused by a series of sparks occurring between the electrically conducting workpiece and a tool with a shape complementary to that of the desired shape of the final product, in the presence of a suitable dielectric fluid [18–21]. Special current generators are used to produce pulsed DC currents to vaporize and melt conductive material away instead of mechanically shearing a chip, as in conventional machining.

WSEM also known as *wire electric discharge machining (WEDM)* is a derivative of *SEM* which utilizes a thin wire as a general purpose tool unlike *SEM or EDM*. Owing to the use of thin wire as a tool, *WSEM* uses a dielectric with low dielectric strength and smaller break down voltage. Deionized water is therefore the most commonly used dielectric due to its low viscosity and rapid cooling rate. To avoid wire breakage contact between the wire and workpiece should be avoided. This implies that WSEM employs a smaller interelectrode gap (IEG), lower voltage and current, shorter pulse-on times, longer pulse-off times (i.e., very high pulse frequency) when compared to *SEM*. Reduced electrode wear, lower energy consumption, and independency from complicated electrode fabrication are some of the advantages of *WSEM* over *SEM* [18–20]. In 1969 the Swiss company “AGIE” manufactured the world’s first WSEM machine with simple features and with limitations of wire material to be copper and brass only. Earlier machines were slow with limited machining ability but as the technology matured the overall ability have improved significantly to meet the requirements of various manufacturing needs. Nowadays, most of the WSEM machines are CNC which helps in improving the efficiency, accuracy, and repeatability.

Micro-SEM uses an electrode with microfeatures to manufacture its mirror image on the workpiece. This requires submicron machine movement resolution to obtain acceptable results. Similarly, in micro-WSEM a micro-size wire (i.e., diameter less than 100 μm) is used to cut the workpiece that is also mounted on an accurate submicron resolution movement table. During micro-SEM the pulse generator may be controlled to produce pulses with durations ranging in length between a few nanoseconds to a few microseconds to control the extent of material removal [18–20]. Considerable research has been done on various aspects of WSEM and micro-WSEM and they have been recognized as superior and sustainable alternatives to the conventional manufacturing processes to manufacture small quantities of precision gears with unique shapes. Therefore most of the following discussion on spark erosion-based machining of gears will focus on gear manufacturing by WSEM.

4.1.3.2 Manufacturing of Gears by Wire Spark Erosion Machining Processes

4.1.3.2.1 Working Principle and Significant Process Parameters

The WSEM process removes workpiece gear material through a series of a high frequency (MHz) spark discharges. Each discharge removes a small volume of workpiece material by melting and vaporization leaving small craters on the work surface. The volume removed by a single spark may be in the range of 10^{-6}–10^{-4} mm^3 but may typically occur about 10,000 times per second [18,19]. Fig. 4.5 depicts an actual photograph of a typical wire-SEM process of gears.

FIGURE 4.5 Wire spark erosion machining of gears.

The process commences with the gear blank being mounted and clamped on the main worktable of the WSEM machine. The worktable is typically CNC in both the X and Y axes in steps of a few microns by means of a stepper motor. A traveling wire is continuously fed from a wire feed spool that passes through the gear blank towards the waste wire spool box. During its travel path the wire is kept under tension by a pair of wire guides which are fixed on the lower and upper sides of the gear blank that maintains a constant gap between the wire and the gear blank. As the material removal or machining proceeds, the worktable (containing the mounted gear blank) is moved along a predetermined path according to the geometry of the gear which is stored as a tool path program in the CNC controller of the WSEM machine. During machining the cutting zone is continuously flushed with deionized water as dielectric. An ion exchange resin is used in the dielectric distribution system to prevent an increase in conductivity and therefore to maintain a constant water conductivity. Significant SEM and WSEM process parameters are [18–21]:

1. *Pulse-on time* or *pulse duration*: The time interval during which the spark (electron discharge) occurs between electrode (wire) and the workpiece once the break down voltage of the dielectric is reached causing its ionization. Consequently the spark then causes erosion of the workpiece material. The longer the pulse-on time, the longer the spark is sustained and the material removal rate increases. However, the resulting craters are broader and deeper which result in a deterioration of the surface finish.
2. *Pulse-off time or pulse interval*: The time duration between consecutive sparks during which there is no current supply to the electrodes and deionization of dielectric takes place. The dielectric also flushes the machining debris from the IEG (machining gap) during this time. Reduced pulse-off times may cause wire breakage and increases the surface roughness of the machined surface due to inadequate flushing. Excessively long pulse-off time increases the machining time and the forces on the wire generated by the dielectric flushing.
3. *Duty cycle*: The ratio of pulse-on time to the cycle time (i.e., sum of pulse-on time and pulse-off time). Most of the SEM and WSEM machines have provision to set it to different values.
4. *Pulse frequency*: The reciprocal of cycle time. Of the four parameters related to the pulsed power supply, certain SEM and WSEM machines have provision to set the duty cycle and pulse frequency while other machines control the pulse by setting pulse-on time and pulse-off time. Most of the manufacturers of SEM and WSEM machines use the former option.
5. *Spark gap voltage*: It is the reference voltage for the actual gap between the workpiece and the electrode/wire for the spark to occur. The machine senses the actual gap and voltage during the machining and attempts to

maintain them to not allow the gap to increase to such an extent that the spontaneous electric discharging stops.

6. *Peak current*: Peak current is the maximum value of the current passing through the electrodes for the given pulse-on time. An increase in its value will increase the pulse discharge energy which in turn may improve the machining rate further. The gap conditions may become unstable for higher peak currents leading to the deterioration of the surface finish.
7. *Tool or wire feed rate:* For WSEM, it is the rate at which the wire is fed through the wire guides, while for SEM it is the rate of downward movement (i.e., plunging) of the tool electrode to shape the required geometry into the blank/workpiece. Frequent wire breakage occurs at lower wire feed rates which deteriorates the surface finish and causes interruptions during machining (i.e., lower productivity).
8. *Wire tension:* It is the tensile load in the wire as it is continuously fed between the wire guides that are used to keep the wire straight between said guides. Unacceptably low wire tension results in wire breakage, dimensional and geometric inaccuracy, and poor machined edge definition.
9. *Dielectric pressure*: It is the fluid pressure of the dielectric at which it is supplied to IEG for flushing the machining debris. Elevated pulse power and machining of a thick workpiece requires higher dielectric pressure whereas for thin workpieces and for trim cuts a lower dielectric pressure is used.

The material and diameter of the wire and the type, conductivity, and flow rate of the dielectric are also important parameters that may influence the ability and effectiveness of an individual WSEM machine.

4.1.3.2.2 Mechanism of Material Removal

Any electrically conductive material irrespective of its hardness and melting point can be processed by SEM/WSEM to manufacture gears, gear cutting tools, ratchet wheels, and splines. A gear of a particular size can be manufactured by SEM and WSEM by using an appropriate size wire and/or electrode. The quality of the gear manufactured by SEM and WSEM is basically affected by the path tracing ability, accuracy of the machine, and appropriate process parameters combinations [22,23]. A schematic representation of mechanism of SEM of a gear is shown in Fig. 4.6. When a DC voltage is applied across the electrodes an emission of electrons (known as primary electrons) from the cathode towards the anode occurs which collides with the molecules of the dielectric. When the applied DC voltage reaches the break down voltage of the dielectric, ionization of the dielectric molecules takes place releasing secondary electrons and forming an ionized column at the closest point between the electrode and the gear blank (Fig. 4.6A). This increases the electric field in the IEG. After the break down of the dielectric

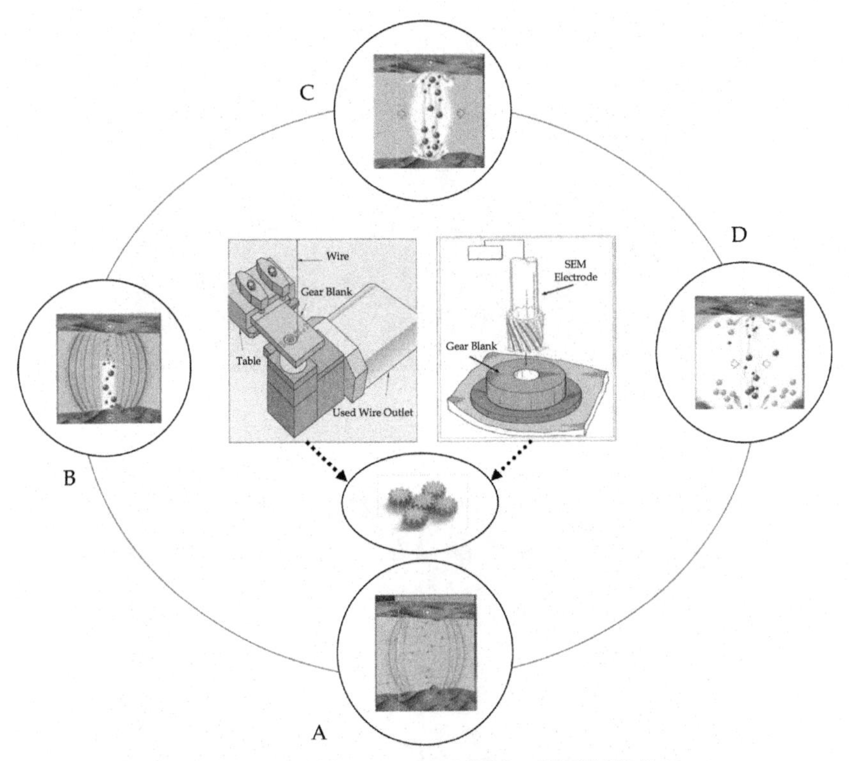

FIGURE 4.6 Mechanism of machining of gears by SEM and WSEM [24].

the voltage falls and the current rises abruptly due to the ionization of the dielectric and formation of a plasma channel between the electrode and the gear blank. This elevated current continues to further ionize the channel and a powerful magnetic field is generated (Fig. 4.6B). This magnetic field compresses the ionized channel and results in localized heating. Even with discharges of short duration the temperature of the electrodes can rise to such an extent that the gear blank material melts locally because the kinetic energy associated with the electrons are transformed into heat. The high energy density erodes a part of the material from both the electrode/wire and gear blank by local melting and vaporization (Fig. 4.6C). At the end of the discharge, current and voltage are shut down (Fig. 4.6D). The plasma implodes under the pressure imposed by the surrounding dielectric. Consequently the molten metal pool is taken up into the dielectric, leaving a small crater at the gear tooth surface. This cycle is repeated until the required amount of material to be removed or the prescribed geometry is realized.

The kinematics of the machine is CNC based and usually controlled by dedicated CAM software. The CNC programing consists of a set of commands to plunge a gear shaped electrode into the workpiece along the Z-axis

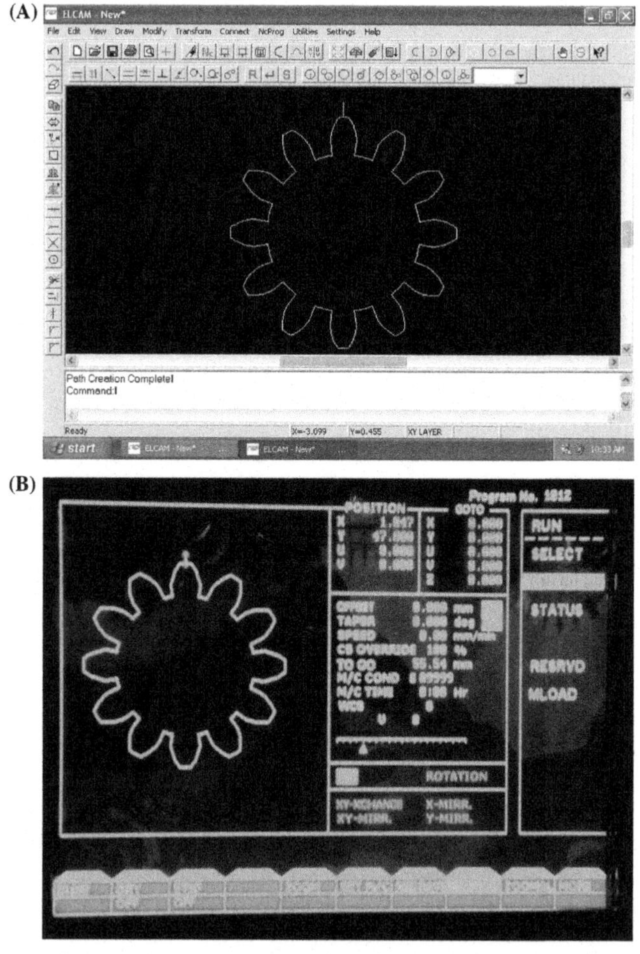

FIGURE 4.7 (A) Gear geometry defined in CAM software; (B) gear geometry representing wire movement after transferring to the WSEM machine tool.

for appropriate process parameters for SEM, whereas a dedicated gear cutting path is traced through the workpiece during WSEM. Fig. 4.7 depicts the geometry of the gear profile defined in the CAM software and the path to be followed by the wire that is then converted in terms of G and M codes to machine a gear. The compensation for electrode size (i.e., wire diameter) and machining overcuts are also specified.

4.1.3.3 Recent Investigations

Published research from the late 1990s to the present amply demonstrates the ability and perceived superiority of SEM-based processes for the manufacturing of gears especially those of miniature type. Probably the earliest work was conducted in 1994 in which a microinvolute spur gear of 0.28 mm outside diameter was fabricated by Hori and Murata [25] using micro-WSEM process with a tungsten wire of 25 μm diameter. It gave a burr-free uniform involute tooth profile with less than 1 μm profile error as demonstrated in the postfabrication metrological investigations and scanned electron micrograph study. Thereafter, in 1997, Suzumori and Hori [26] developed and successfully tested a prototype wobble motor equipped with stator and rotor with composite (involute and arc) teeth profiles fabricated by micro-WSEM for high torque and low load applications. Using micro-SEM, in 2000, Takeuchi et al. [27] fabricated a microplanetary gear system from SKS3 tool steel and WC–Ni–Cr super hard alloy for a chain-type self-propelled micromachine used in power plants. Benavides et al. [28] (Manufacturing Science and Technology Centre at Sandia National Laboratories in the United States) employed micro-WSEM to fabricate a mesosized ratchet wheel from a wide range of materials including 304 L austenitic stainless steel, nitronic 60, beryllium copper, and titanium with submicron level surface finish, burr-less edges and profiles, minimum recast layer, and consistent microgeometry.

Recently, an in-depth study on near-net shape manufacturing of gears by WSEM has been conducted in the Gear Research Lab at the Indian Institute of Technology Indore, India [29]. This study encompasses the evaluation of the effects of WSEM parameters on gear quality and surface integrity, optimizing the WSEM process to enhance performance characteristics and the service life of the gears, and process performance comparison with conventional methods of gear manufacturing. It was specifically focused on manufacturing external spur miniature gears of brass (ASTM 858). This study reveals that active control of WSEM process parameters is required to obtain the desired fit for purpose gear quality. Gears of desired geometry and surface properties can be manufactured by ensuring minimum wire lag (i.e., deviation of wire from its intended path) and nonviolent spark discharges at optimal parameter settings without the aid of any postfinishing operation [30,31]. WSEM can successfully manufacture gears with quality up to DIN-5 standard which is significantly superior to the quality of gears manufactured by other conventional processes and on par with gears finished by grinding, honing, and shaving. Moreover, manufacturing with active control of the process parameters may result in uniform flank topography, burr-free teeth profile, acceptable surface finish, good microstructural aspects, and negligible recast layer on gear teeth [29]. This ensures improved operating performance and enhanced service life of gears. Earlier, it was not possible to machine helical and bevel type typical gear profiles, but the technological advancements in the form of 5 axis machining ability, improved machine kinematics and accurate repeatability, automatic work changer systems, increased taper angle ability, downsizing of the wire diameter, reliable, and

Left-hand external helical gears from SS304

External spur gear with hub from aluminium

External spur gears from aluminium, copper, brass, and SS304

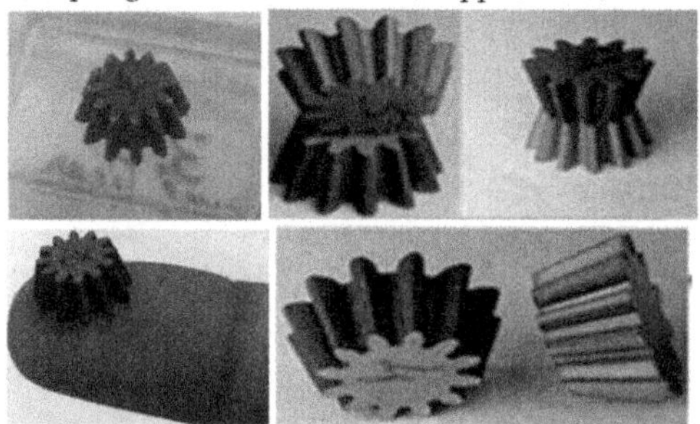
Straight bevel gears from brass and SS304

FIGURE 4.8 High-quality miniature cylindrical and conical gears manufactured by wire-SEM at Indian Institute of Technology Indore, India.

consistent automatic wire threading ability, features to facilitate long hours of unattended machining, etc. has permitted rapid machining of complex gear shapes with high accuracy and without significant human (operator) intervention. Fig. 4.8 displays typical examples of high-quality cylindrical and conical miniature gears manufactured by WSEM from different materials (at IIT Indore, India).

4.1.3.4 Advantages, Capabilities, and Limitations

SEM and its variants have the following significant advantages over conventional processes of gear manufacturing [29–32]:

- The single-stage manufacturing of gears significantly limits the extent of the process chain;
- Reduced setup time, cycle time, and overall gear manufacture process time;
- Very less loss of gear material during the manufacturing is highly material efficient;

- Lower consumption of power, material, cutting tools, and fluids;
- Better surface integrity and manufacturing quality of gears;
- Cost of producing an acceptable quality gear is significantly lower than that of other conventional processes of gear manufacturing;
- Improved ecological prospects.

Major capabilities of SEM-based machining processes with regards to manufacturing of gears are [32–35]:

- Tooth corrections and modifications including root alteration, tip relief, and crowning can be done;
- Machining of large gears can be done unattended during an overnight shift;
- Able to manufacture the internal or external gears of any shape (helical, bevel, and noncircular, etc.) with accuracy equivalent to precision gear finishing processes such as gear grinding, honing, and lapping;
- Hardness of the workpiece material is irrelevant;
- Lack of direct contact between the workpiece and wire electrode ensures no induced mechanical stress;
- Average roughness (R_a) value as low as 0.05 μm can be achieved by micro-SEM and micro-WSEM processes;
- Tolerances as close as ±0.5 μm can be attained by micro-SEM/WSEM processes;
- Gears with module size ranges of 10–30 μm can be manufactured by micro-WSEM equipped with 10 μm diameter wire (currently the smallest practical wire diameter).

Despite offering many advantages and ability for gear manufacturing, SEM-based processes suffer from the following significant limitations [29]:

- High capital cost;
- Not suitable for mass production;
- Not practical to fabricate very large mill gears;
- Gears made of electrically nonconducting materials such as plastics, and most composites cannot be manufactured by SEM/WSEM;
- Formation of a recast layer particularly at elevated discharge energies is sometimes a problem in SEM-based machining, which can be eliminated by controlling the pulse parameters of the process;
- Wire lag and nonoptimum discharging (sparking) may adversely affect the gear quality and surface integrity. Therefore machining at the optimal process parameter settings is desirable.

In essence, to manufacture gears of excellent quality and good surface integrity, an appropriate machine tool with optimum combination of process parameters is essential for effective SEM and WSEM. No doubt that these processes are able to manufacture quality gears at significantly low cost and with better ecological prospects compared to the conventional processes of gear manufacturing. This technology may yet still develop to be a superior sustainable alternate to conventional methods to manufacture quality gears.

4.2 ADDITIVE OR ACCRETION PROCESSES

Additive or accretion type processes manufacture gears or components by utilizing a bottom-up approach by processing material either in liquid or powdered state unlike subtractive manufacturing processes that is based on a top-down approach and process the material in the solid state. Selected modern additive manufacturing processes for gears including metal injection molding (MIM), injection compression molding (ICM), micropowder injection molding (μ-PIM), and additive layer manufacturing (ALM) are described in the following sections. The ALM processes include stereolithography (SLA), selective laser sintering (SLS), 3D-Printing (3DP), and fused deposition modeling (FDM).

4.2.1 Metal Injection Molding

MIM hybridizes injection molding with powder metallurgy to manufacture net-shaped or near-net-shaped gears and other complex meso- and micro-sized products. Being a hybrid of additive type manufacturing processes, it is an economical alternative to many conventional manufacturing processes such as investment casting, powder metallurgy, and stamping.

4.2.1.1 Working Principle

MIM utilizes a mixture of metallic powder and organic binder injected into a split die under high pressure rather than letting it to flow under gravity [36]. MIM typically consists of a sequence of five activities (see Fig. 4.9):

- *Preparation of raw material*: Fine metallic powder (particle size <20 μm) is mixed with an organic binder such as wax or some polymer in a ratio of 3:2 to the raw material which is placed in mixer and heated to a temperature which melts the organic binders. Mechanical mixing is then conducted to uniformly coat the powder particles in the binder, which is then followed by cooling to form free flowing pellets that can be used as raw material to the injection molding machine.
- *Injection molding*: The raw material (in the form of pellets or granules) is fed to the heated cylinder of a plastic injection molding machine from its hopper. Heating of the raw material melts the binder at approximately 200°C. The raw material is then fed forwards towards a split-die chamber or an appropriate mold cavity by a rotating screw system. As the pressure builds up at the entrance of the mold cavity, the rotating screw retracts

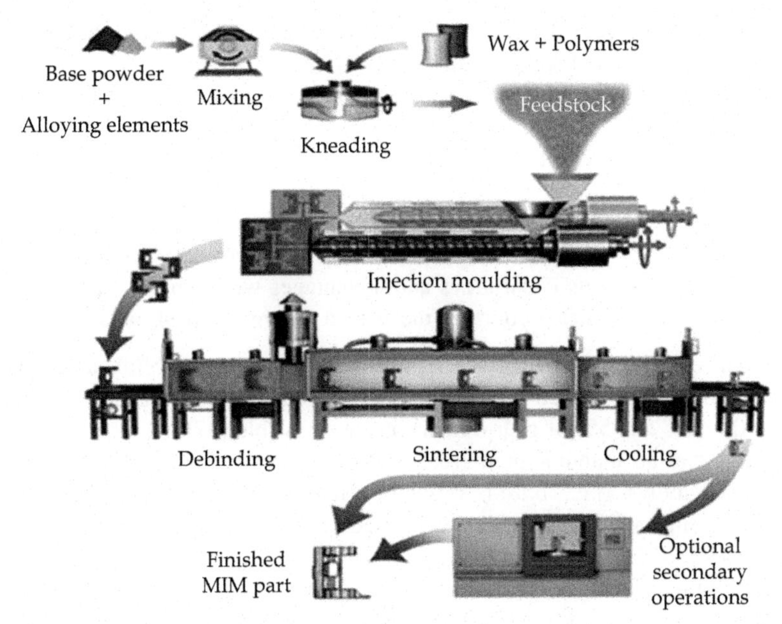

FIGURE 4.9 Sequence of activities involved in the metal injection molding process. *From N. K. Jain, S.K. Chaubey, Review of miniature gear manufacturing, in: M.S.J. Hashmi (Ed.), Comprehensive Materials Finishing, vol. 1, Elsevier, Oxford, UK, 2016, pp. 504–538, doi:10.1016/B978-0-12-803581-8.09159-1. Elsevier © 2017. Reprinted with permission.*

linearly under pressure to a predetermined distance which controls the volume of material to be injected.

The rotation of the screw is then stopped and is displaced linearly forwards hydraulically to inject the raw material into the mold cavity. The injected part is cooled and extracted from the die. In this state the molded gear is referred to as in its "green state." It is typically 20% larger than the final size to compensate for shrinkage during the sintering process. The type of gear material determines the exact value of shrinkage allowance.

- *Debinding*: Removes almost 80% of the binders from the molded part either by using a catalyst or heating process or a closed loop solvent or combination to yield the so-called *brown gear* which is semiporous and fragile but without any change in dimensions. Debinding time depends on the gear thickness.
- *Sintering*: It involves heating the brown gear in a vacuum furnace to a temperature below the melting point of the gear material but still sufficiently high that the metallic particles will fuse thereby reducing the porosity and eliminating the remaining binder. Sintering provides high dimensional accuracy and shrinks the brown gear by 15–25%. Density

of the sintered part is generally 95–99% of the density of the gear material. Debinding and sintering may take 24–36 h.

- *Postsintering*: The final properties and surface characteristics of the gear can be further improved by utilizing some property enhancing process such as case hardening, cold working, coating, etc.

4.2.1.2 Capabilities, Advantages, Limitations, and Applications

MIM has the following capabilities and advantages which make it a versatile process for manufacturing complex mesosized and microsized parts:

- Produces an excellent surface quality with good repeatability and close tolerances. Net-shaped or near-net shaped gears with tolerances between ± 0.3 and ± 0.5% of gear dimensions are possible [37].
- Gears can be manufactured from various materials including aluminum, copper alloys (i.e., brass), refractory metals, ferrous alloys, titanium alloys, super alloys, ceramics, cemented carbides, and metal matrix composites [38].
- Economical for small gears weighting less than approximately 100 g, but can also be used for microgear weighting less than 0.1 g and larger gears up to approximately 250 g.
- This process is most suited to components with thicknesses less than 6 mm although thicker components can also be manufactured readily.
- Cost-effective when compared to conventional processes for manufacturing combined components (i.e., gear and shaft) which reduces the cost of individual parts and assemblies.
- This process has high scalability because production volume from several thousand to millions of parts annually can be manufactured economically.
- It results in less material wastage and faster production when compared to investment casting. Cost of gears manufactured by MIM may reduce up to 50% compared to investment casting and/or CNC machining.

Despite possessing some unique capabilities and advantages, MIM does display certain limitations:

- High shrinkage of material ranging from 25% to 30% during the sintering process;
- Mostly suitable to manufacture small gears;
- Higher lead time than injection molding;
- MIM process is more expensive than injection molding due to multiple steps and higher material and mold cost.

Some examples of gear manufacture by MIM include the manufacture of complex meso and microsized gears for various industrial and commercial applications [39, 40].

4.2.2 Injection Compression Molding

ICM is a combination of rapid but incomplete charging of the partially closed mold before it is completely closed and uniform compression is applied. It may be considered as a modified injection molding process with an additional compression system. Although this hybridization may enhance the production rate, but it also increases complexity and make its control more difficult. The thickness of the gear or target part should be less than the depth of the mold cavity so that the molten raw material is able to completely fill all extremities of the mold cavity. ICM is also referred to as compressive-filling, hybrid molding, stamping, or even as coining process [41].

4.2.2.1 Working Principle

A schematic of the working principle of ICM is depicted in Fig. 4.10. The manufacturing of a gear or part essentially occurs in two steps namely injection and compression involving the following sequence of events [42]:

- *Injection*: This step involves (1) displacement of the movable portion of the mold cavity towards the fixed platen; and (2) filling the mold cavity with the required amount of the molten raw material with the injection molding machine from raw material extracted from its hopper. Initially

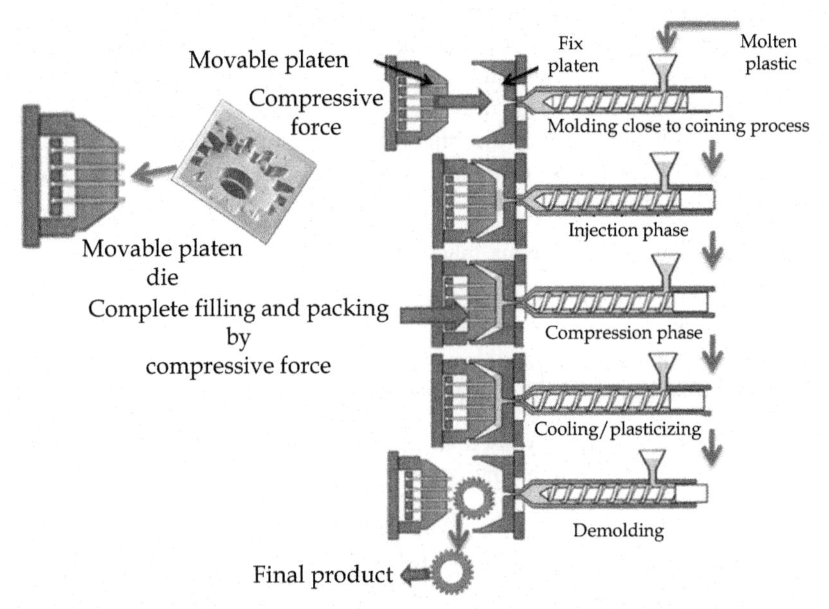

FIGURE 4.10 Schematic of the working principle of injection compression molding. *From N. K. Jain, S.K. Chaubey, Review of miniature gear manufacturing, in: M.S.J. Hashmi (Ed.), Comprehensive Materials Finishing, vol. 1, Elsevier, Oxford, UK, 2016, pp. 504–538, doi:10.1016/B978-0-12-803581-8.09159-1. Elsevier © 2017. Reprinted with permission.*

the mold cavity is kept partially open so that the raw material may flow uninhibited. After the completion of the injection phase, all inlet gates are closed and the filling stage is complete.

- *Compression*: During the compression step the mold is fully closed to obtain the final configuration of the gear/part. This step involves. (3) Filling of all the unfilled corners of the mold cavity by compressing the molten raw material. It also results in a uniform packing pressure over the entire mold cavity and helps in achieving the final complex geometry of the gear; (4) cooling or plasticizing the gear mold; and (5) opening of the mold cavity and subsequent ejection of the molded gear.

4.2.2.2 Capabilities, Advantages, Limitations, and Applications

ICM affords the following capabilities and advantages [41,42]:

- Residual stresses are lower due to the use of lower clamping forces and more homogeneous physical properties (i.e., less variation in density) due to uniform packing and filling of all the corners of the mold cavity;
- Large volume manufacture of complex plastic gears which cannot be manufactured by the conventional injection molding process;
- Enhanced dimensional accuracy and stability due to uniform distribution of the raw material and elimination of warpage;
- This process may compensate for uneven shrinkage;
- Lower total manufacturing time due to lower injection pressure and clamping force, reduced holding pressure time, minimal molecular orientation, and alignment needs to occur;
- Process is flexible in that the injection and compression stages may be sequenced in various different ways and may also include post-compressive if required.

ICM may have the following inherent limitations:

- Overfilling of the mold causes wastage of the raw material;
- High cost of the mold and machine;
- Flash control and removal;
- Process control is difficult and complex;

ICM process is generally used to manufacture thin and lightweight, but strong and stiff gears of high quality.

4.2.3 Micropowder Injection Molding

μ-PIM is an additive type advanced manufacturing process that hybridizes powder metallurgy with microinjection molding. It was developed specially to manufacture microcomponents with features less than 50 μm size. It has evolved as a reliable process to manufacture microgears made of ceramics where powders are readily available in submicron size. Moreover, submicron size ceramic powders are safer and easier to handle than the submicron size metallic powders that may be pyrophoric [43].

4.2.3.1 Working Principle

Micro-PIM utilizes higher injection pressures than typically used in the conventional injection molding process. Manufacturing a microgear by μ-PIM process involves following steps as illustrated in Fig. 4.11 [44]:

- *Preparation of raw material*: Raw material for μ-PIM process is prepared by preparing appropriate shaped pellets of a mixture comprising of micro/nanosized metallic/ceramic powder and an organic binder by extruder or mixer. The role of the organic binder is to act as the carrier medium for the metallic/ceramic powders and is removed in the subsequent step. A variotherm process is generally used to avoid risk of solidification of the prepared raw material which may occur due to (1) its high thermal conductivity and (2) small dimensions of the cavity.
- *Microinjection molding*: It implies feeding the raw material into the mold cavity to obtain the desired shaped gear. In this state it is referred to as "green" as it requires addition steps for completion. This is similar to

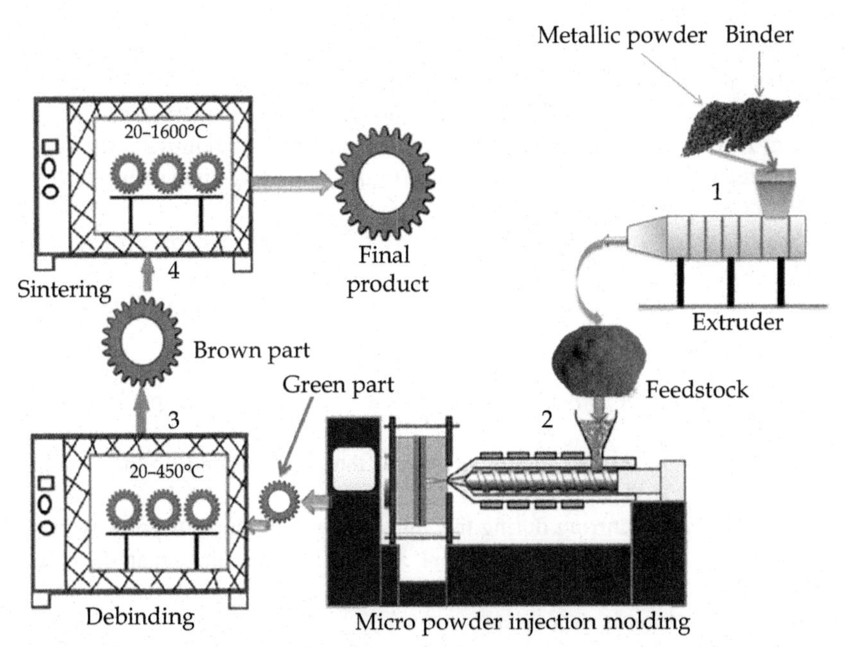

FIGURE 4.11 Schematic of microgear manufacture by the μ-PIM. *From N.K. Jain, S.K. Chaubey, Review of miniature gear manufacturing, in: M.S.J. Hashmi (Ed.), Comprehensive Materials Finishing, vol. 1, Elsevier, Oxford, UK, 2016, pp. 504–538, doi:10.1016/B978-0-12-803581-8.09159-1. Elsevier © 2017. Reprinted with permission.*

conventional plastic injection molding except some changes are made to address the compressibility and viscosity of the raw material. The gear is made slightly oversized to allow for shrinkage during sintering.

- *Debinding*: Removal of the organic binder from the green gear to obtain the so-called *brown gear*. The debinding process can be performed in three ways (1) eliminating the organic binder by thermal degradation along in an appropriate temperature controlled environment; (2) using supercritical carbon dioxide in an autoclave operating at more than 57°C and 300 bar pressure; and (3) using a catalyst or solvent [45].
- *Sintering*: Sintering is typically done in a tube furnace to achieve better mechanical properties of the *brown gear*. Air is used for sintering of ceramic microgears whereas a reducing H_2 atmosphere is used for microgear/parts in metallic materials. Sintering may result in shrinkage of the brown gear by between 20% and 30%. The actual value depends on the composition of the raw material [46]. The final properties of the gear are generally similar than that of the raw material.

4.2.3.2 Capabilities, Advantages, Limitations, and Applications

μ-PIM affords the following capabilities and advantages [47]:

- Dimensionally accurate and complex microgears and 3D microparts can be manufactured from ceramics and metallic materials. It offers the design flexibility of injection molding and the advantages of powder metallurgy;
- It has ability to address various limitations of MIM of microgears including: incomplete filling of narrow cavities, difficulty in removing the fragile green compacts and distortion due to the debinding and sintering;
- Reduced raw material wastage;
- Smaller sized and prototype gears may be manufactured by PIM at low pressure (1–10 bar) and utilizing low-viscosity paraffin or wax as binder;
- It is economical and cost-effective for mass production of microgears.

Micro-PIM has the following limitations:

- Use of only fine powders with particle sizes less than approximately 5 μm;
- Surface roughness and minimum size of the gears depend on the powder size;
- Grain growth occurring during the sintering and presence of porosity may adversely affect the mechanical properties of the manufactured microgears;
- Mold fabrication is expensive;
- Longer manufacturing time due to debinding and sintering.

The μ-PIM process can be used to manufacture various microparts including stepped microgears, microgears of any shape and profile, and microgearboxes for applications including clocks, cameras, electronic, biomedical devices, etc.

4.2.4 Additive Layer Manufacturing Processes

4.2.4.1 Introduction, History, and Development

ALM is an advanced manufacturing technology which uses a bottom-up approach to manufacture a complex part or even an assembly from its CAD model by layered deposition of the material. ALM can be used to manufacture either (1) new net-shaped or near net-shaped complex component or assemblies; and/or (2) to add delicate features to an existing component. ALM technology offers many advantages over conventional manufacturing processes based on a top-down approach. Some significant advantages are: (1) it is material and energy efficient due to significant reduction in scrap; (2) it may be faster than conventional manufacturing processes; (3) no detailed working drawings of the components or parts are required; (4) it eliminates the task of procuring materials in specific size and shape; (5) it also eliminates the task of the production/process planning specific to part, machine, and manpower; (6) no intermittent quality checks are required; and (7) human intervention errors are limited.

These unique set of advantages implies that ALM may be significantly more environment-friendly and sustainable as a manufacturing process. Three well-known significant applications of ALM are: (1) rapid prototyping (RP) which involves near instantaneous production of patterns and/or products for design and evaluation purposes; (2) rapid tooling (RT) which involves making tools, dies, molds, EDM tools, etc.; and (3) rapid manufacturing (RM) which implies making real-life or fully functional products. Several ALM processes have been developed since the 1980s. These ALM processes can be classified into the following three categories according the state of the raw material used:

- *Liquid-based ALM processes*: Commences with a material in the liquid state that cures to a solid state. Examples are SLA apparatus, solid ground curing; solid creation system, solid object ultraviolet (UV)-laser plotter, etc.
- *Solid-based ALM processes*: Raw material is used in the form of wire, pellets, laminates, or rolls. Examples are: FDM, laminated object manufacturing, multijet modeling, selective adhesive and hot press, paper lamination technology, RP system, laser metal wire deposition, etc.
- *Powder-based ALM processes*: Raw material in the powder form is used. Examples are: fused deposition of ceramics, ballistic particle manufacturing, three dimensional printing (3DP), multiphase jet solidification, direct shell production casting, SLS, direct metal laser sintering, direct laser

forming, laser rapid manufacturing, laser engineered net shaping, direct metal deposition, shaped metal deposition, electron beam melting, etc.

Most of the ALM processes are able to manufacture components from various advanced and novel materials including certain plastics, fiber reinforced composites, shape-memory alloys, and functionally graded materials. Certain processes can manufacture components from ceramics also [48–50].

Rapid and continuous development of various ALM processes with their different characteristics has revolutionized the manufacturing of sophisticated engineered parts, devices, and systems in general and of miniaturized parts in particular. This also has opened up new facets in gear manufacturing. The gears manufactured by ALM find applications in numerous areas viz., robots, micromotors, biomedical devices, scientific instruments, harmonic-drives, printers, and other electronic gadgets, to name just a few. In short, ALM is a cost-efficient and sustainable technology to produce good quality, lightweight, and quite gears of high reliability.

The concept of ALM was developed in the 1980s. In 1981, Hideo Kodama of Nagoya Municipal Industrial Research Institute invented two ALM fabricating processes for 3D plastic model manufacture. It used a photo-hardening polymer, where the appropriate area to be hardened is exposed to UV scanning via a mask pattern [51,52]. Interestingly, in 1984 parallel patents were filed by Murutani (Japan), Andre et al. (France), Masters (United States), and Chuck Hull (United States) describing a similar concept of layer-by-layer fabrication of 3D-objects [49]. The prototype developed by Chuck Hull of 3D System Corporation and based on a process known as SLA (in which layers are added by curing photopolymers with UV light lasers), was recognized as the most influential and gave further impetus to the commercialization of 3D ALM systems utilizing SLA. Thereafter in 1986, SLS was patented and followed by FDM and 3DP in 1989 [49]. Since their inception, these techniques have frequently been employed to produce complex 3D-objects including gears and other mechanisms. Budzik [48] and his team have extensively worked on ALM technology to manufacture a variety of gears. A glimpse of their work using different ALM processes for RP of gears is presented in various parts of this section.

4.2.4.2 Additive Layer Manufacturing Processes Steps for Gear Manufacture

The typical steps required for gear manufacturing (or any other object) by ALM are presented in Fig. 4.12. It commences with the creation of a 3D-CAD solid model of the gear. This step is followed by creating a STL format-based file from the CAD model of the gear that contains information as regards to the geometry only and approximating the surfaces with a series of triangular facets in a process referred to as *tessellation*. In

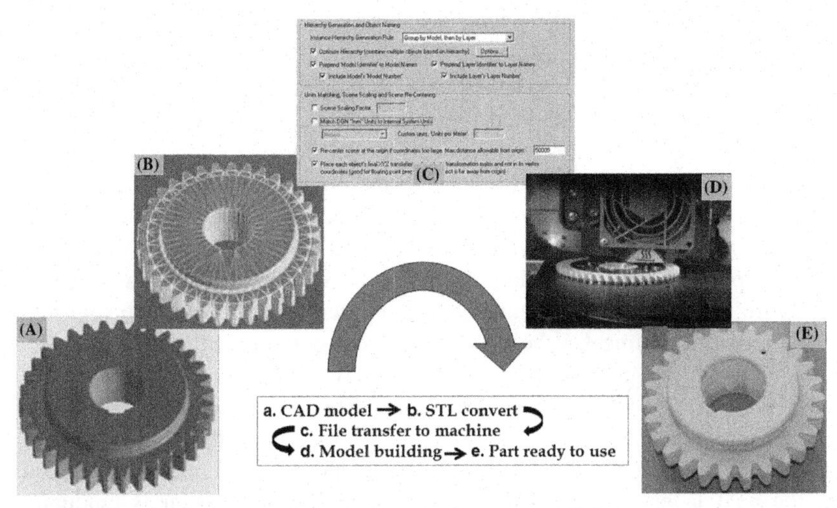

FIGURE 4.12 Typical steps required to manufacture a gear by ALM.

other words the STL file simply converts the CAD model of the gear/part into 3D meshes of triangular elements. Sometimes, CAD software produces errors in the STL files that have to be repaired before further processing. The next step is slicing, i.e., intersecting the CAD model with a plane in order to determine 2D contours and to define layers. Thereafter the CAD model is converted into a series of thin layers and the corresponding part program file containing instructions tailored to a specific type of ALM machine is generated. The prefinal step is the actual manufacturing of the gear (i.e., physical model) according to the CNC part program. In some instances, some postmanufacturing operations such as cleaning of manufactured gear, removal of support material, and finishing, etc. are required to impart the final shape, finish, and quality to the gear [48–50].

4.2.4.3 Stereolithography

SLA is an ALM process that works by focusing an UV laser to a specific and controlled surface location of a photopolymer resin contained within a vat [49,50]. The UV laser is used for curing the liquid photopolymer on the surface in a layer-by-layer process to obtain the preprogrammed design or shape of the gear. Since, photopolymers are photosensitive under UV light, the resin is solidified and forms a single layer of the desired gear. This process is repeated for each layer of the design until the gear is complete. In other words the technique is based on the process of photo-polymerization, in which a liquid resin is converted into a solid polymer using laser irradiation. A typical SLA system/machine consists of a computer, a vat containing a photosensitive polymer resin, a movable platform or elevator on which the model is built, a laser to irradiate and cure the resin and a dynamic mirror system to direct the laser beam. The basic principle of gear manufacturing by SLA is schematically shown in Fig. 4.13. A CAD 3-D solid model of the gear divided (sliced) into a series of 2D layers of uniform thickness. The set of layers (sliced model information) is used to control a UV laser beam (He–Cd laser with the power ranging from 20 mW to 1 W) and its focusing point via a mirror towards an appropriate location on the surface of the liquid polymer. A CNC part program (typically in terms of appropriate G and M coding) for each 2D sliced layer is then executed to control a motorized x–y stage containing the vat of UV curable solution. The focused scanning UV beam induces local polymerization, i.e., conversion of the liquid monomer to the solid polymer. As a result, a polymer layer is formed according to each 2D layer file and a fraction of the physical 3D part/gear is created. Upon completion of a layer the elevator moves downward to expose a new layer of liquid monomers that may be solidified as the next layer on the platform. Subsequently the part/gear is gradually displaced downwards (along the z-axis) in such a way that the individual hardened layers are adequately joined to make up a uniform part/gear. Synchronization of the x–y scanning and the z-axis motion is the key to manufacture high quality gears. Compensation for the laser beam radius and polymer shrinkage are considered to minimize errors. The resin composition includes photoinitiators, prepolymers, a reactive dilute, and additives [50]. Fig. 4.14 depicts a variety of gears manufactured by Budzik [48] using the SLA.

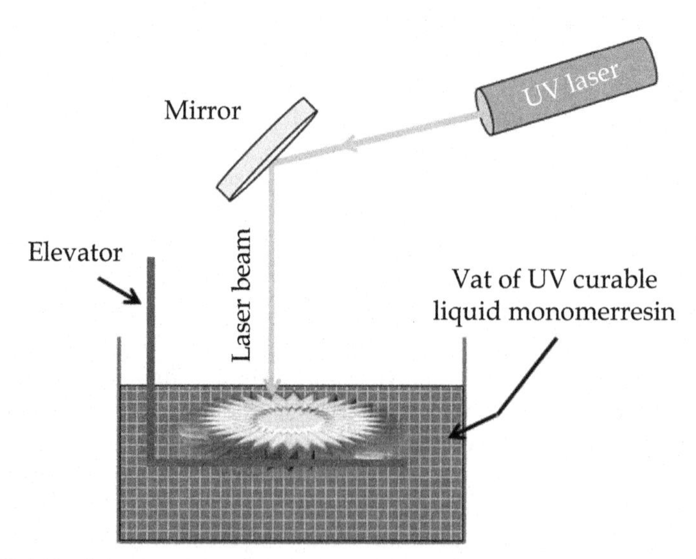

FIGURE 4.13 Schematic of gear manufacture by SLA process.

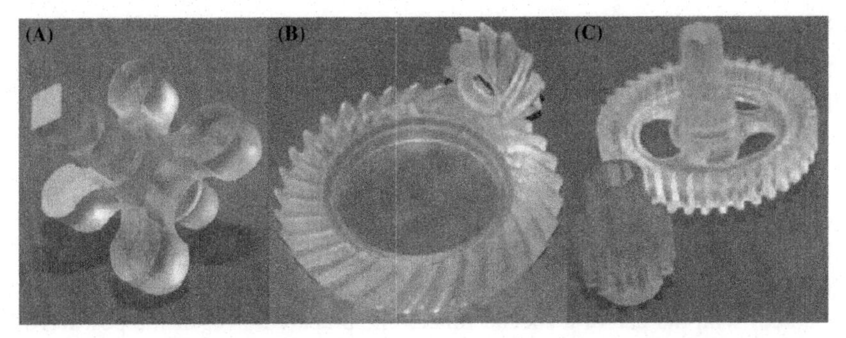

FIGURE 4.14 Different gear wheels manufactured by SLA: (A) gear with cycloidal profile; (B) a set of bevel gears; (C) cylindrical gears. *From G. Budzik, The use of the rapid prototyping method for the manufacture and examination of gear wheels, in: M. Hoque (Ed.), Advanced Applications of Rapid Prototyping Technology in Modern Engineering, InTech, Croatia, 2011, ISBN: 978-953-307-698-0. © Budzik G published under CC BY 3.0 license, available from: http://dx.doi.org/10.5772/22848.*

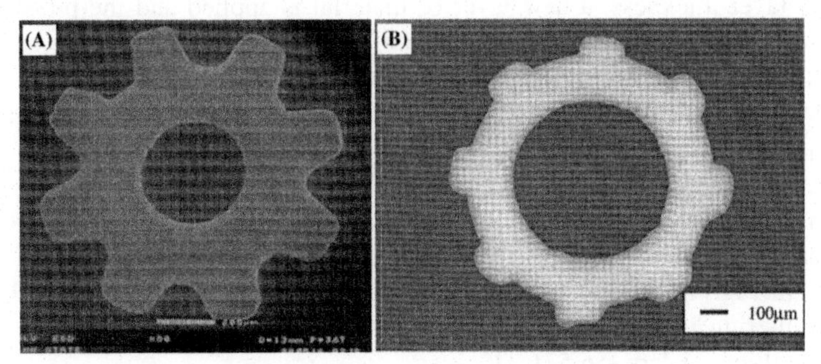

FIGURE 4.15 Microgears (1000 μm diameter) manufactured by μ-SLA from (A) aqueous alumina suspension with solid loading of 33%; and (B) nonaqueous alumina suspension with the solid loading of 33% [53].

Investigations have been conducted to develop a microversion of SLA, i.e., μ-SLA process to manufacture microgears and nanogears as used in microelectromechanical and nanoelectromechanical devices including micromotors, biomedical devices, and microrobots. A high resolution (1.2 μm) micro-SLA system was designed and developed by Zhang et al. [53] that was successfully used to manufacture 100 and 400 μm diameter microgears from a silicon substrate. They also manufactured 400 and 1000 μm diameter microgears from alumina (see Fig. 4.15) using laser power in the range of 5–15 mW [53]. The work conducted by Yoshinari et al. [54] demonstrated the processing of microgears made of zirconia toughened alumina (ZTA) dispersed in acrylic resin. The perceived quality the gear suggested a potential future for micro-SLA for producing ceramic gears to be used in MEMS, biomedical and other precision applications [54].

4.2.4.4 Selective Laser Sintering

SLS was developed at University of Texas, Austin (United States) and subsequently continuously developed with the invention to manufacture complex functional parts made from a wide range of materials. The basic working principle is depicted schematically in Fig. 4.16. Typically an SLS system consists of a powder delivery system, roller, UV laser system, galvanometric mirrors, and working platform. The SLS process begins with the application of a powder layer from a supplementary container. A measured amount of the powder is dosed by a delivery system (piston based) and then spread (distributed) with the use of a special roller. The prepared layer of the powdered material is then locally sintered by a pulsed laser beam. The process is repeated in a layer-by-layer process similar to other layer deposition techniques to build up a part/gear. A high power laser (e.g., carbon dioxide laser, etc.) is used which selectively fuses powdered material or causes the sintering of the powder layer by scanning cross-sections generated from the 3D-CAD model. After each cross-section is scanned, the powder bed is lowered by one layer thickness, a new layer of material is applied and the process is repeated until the entire gear/part is manufactured. Once the complete gear is manufactured the powder which has not been sintered is removed. The bulk powder material in the powder bed is preheated, to reduce the time it takes for the laser to increase the temperature as required [49,55].

Gears made of plastics, metallic powders, mixtures of metallic and ceramic powders, and polymer and ceramic powders can be manufactured by this process. Examples of polyamide and TiAl6V4 alloy powder are presented in Fig. 4.17. Microversion of SLS process referred as selective laser microsintering is a good solution to manufacture microgears from ceramic and metallic powders [56,57].

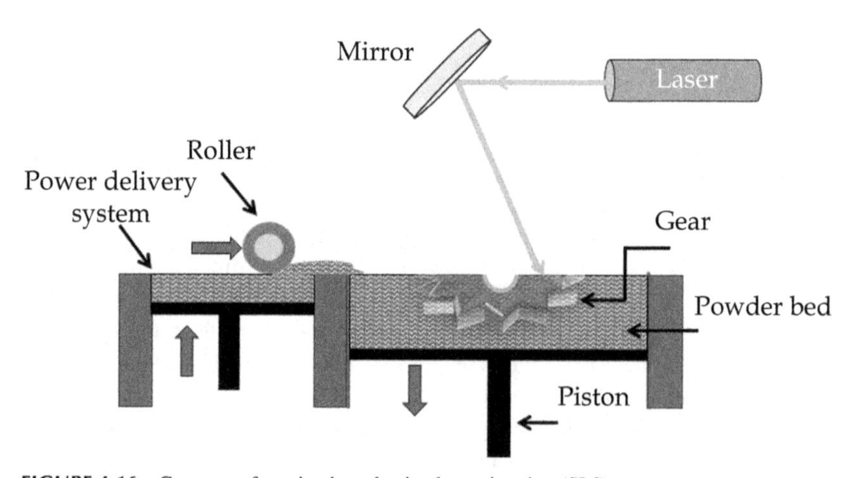

FIGURE 4.16 Gear manufacturing by selective laser sintering (SLS) process.

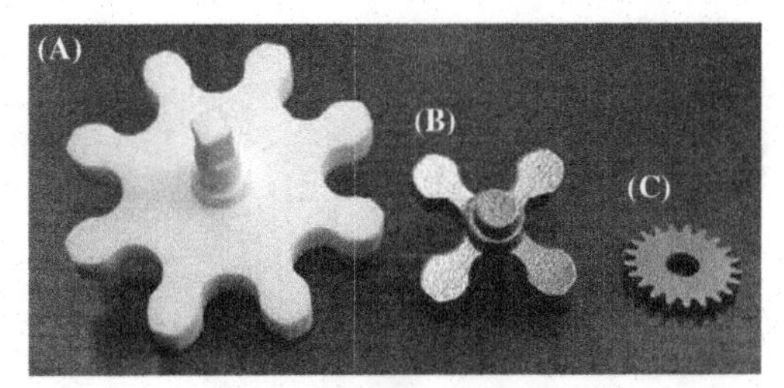

FIGURE 4.17 Gears manufactured by SLS: (A) PA2200 polyamide; (B) and (C) TiAl6V4 titanium alloy. *From G. Budzik, The use of the rapid prototyping method for the manufacture and examination of gear wheels, in: M. Hoque (Ed.), Advanced Applications of Rapid Prototyping Technology in Modern Engineering, InTech, Croatia, 2011, ISBN: 978-953-307-698-0, © Budzik G published under CC BY 3.0 license, available from: http://dx.doi.org/10.5772/22848.*

4.2.4.5 3D-Printing

Three-dimensional printing (3DP) was first developed at the Massachusetts Institute of Technology where after "Z Corporation" obtained its exclusive license. This process works by selectively binding powdered material deposited in layers by means of a binding agent injected through a print head to manufacture a thin cross-section of the 3D part/gear one at a time [58–61]. Owing to the application of binder agent, this method is also known as *binder jet 3D printing or binder jetting*. Fig. 4.18 presents the various parts and working principle of typical binder jet 3D printing process. It comprises a powder delivery system, a roller, an inkjet print head, and a working platform. It commences by depositing a measured amount of the powder on the working platform that is subsequently layered with the help of a special roller. The binding agent is then applied to the prepared powder layer via an inkjet print head in accordance with the defined cross-section of the gear/part. The inkjet print head deposits droplets of binder onto the powder layers and effectively bonds the power particles together for each individual layer. The working platform is lowered by as appropriate for the next layer to be deposited and the cycle is repeated until all the successive layers are formed to complete the gear. Upon completion unbound powder is automatically and/or manually removed in a process called *de-powdering* and may be reused to some extent. The depowdered gear could optionally be subjected to various infiltrates or other treatments to obtain the desired quality [48]. Gears of metals (SS, aluminum, etc.), ceramics, composites, and plastics powders can be manufactured by binder jet type 3DP. Epoxy, wax, and cyanoacrylate glue, etc., are some frequently used binding materials [48,59]. This process makes it possible to "print" geometric models (Fig. 4.19A) as well as molds (Fig. 4.19B) for casting alloys at pouring temperatures of up to 1100°C (Fig. 4.19C).

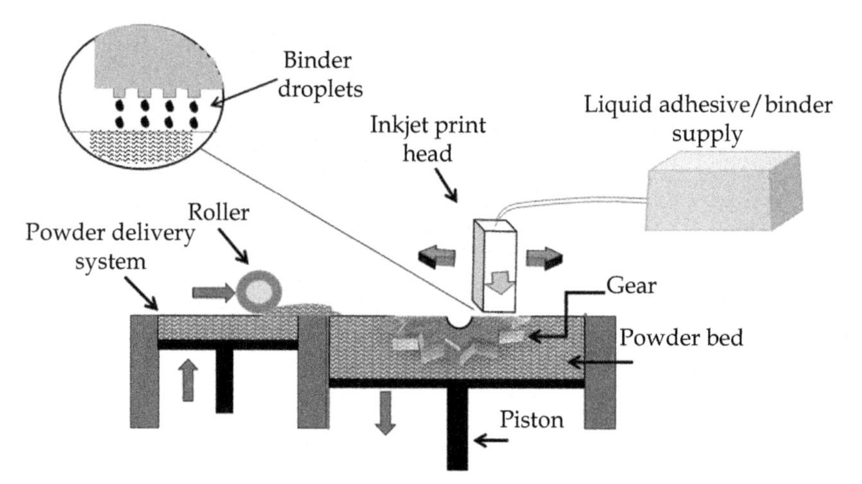

FIGURE 4.18 Schematic of the working principle of binder jet type 3-D printing (3DP).

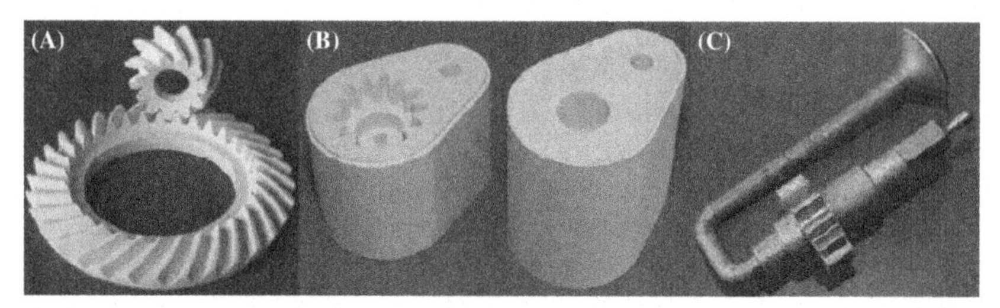

FIGURE 4.19 Parts manufactured by binder jet type 3D-printing: (A) bevel gears; (B) mold for casting gears; (C) Al alloy cast gear produced using the mold. *From G. Budzik, The use of the rapid prototyping method for the manufacture and examination of gear wheels, in: M. Hoque (Ed.), Advanced Applications of Rapid Prototyping Technology in Modern Engineering, InTech, Croatia, 2011, ISBN: 978-953-307-698-0, © Budzik G published under CC BY 3.0 license, available from: http://dx.doi.org/10.5772/22848.*

PolyJet printing is a 3DP technology developed by *Objet Geometries Ltd* which is now a part of Stratasys Inc. The principle of the *PolyJet* process is based on jetting a liquid photopolymer onto a build platform, its immediate curing by UV light and associated solidification [58–61]. A wide range of plastics including acrylonitrile butadiene styrene (ABS), acrylate, polypropylene, and other photopolymers can be formed into gear shapes by this process. Polyjet machines utilize inkjet print heads to spray a liquid photopolymer in ultra-thin layers with thicknesses between 16 and 30 μm onto a build platform that is able to displace vertically (Z-axis). Polymer layers are applied one by one via the print head onto the working platform (x, y). Generally, two materials namely the base material and support material are used to manufacture a gear/part. After jetting, each photopolymer layer is immediately cured by UV light emitted by a lamp integrated with the print head and is subsequently solidified which allows adding layers on top of each other until the complete gear is manufactured. It produces fully cured gears that can be handled and used immediately without any postcuring process. Sometimes, a gel-like material is used to support complicated geometry associated with gear/part shapes that is typically easily removed by hand and water jetting. Typically gears up to 150 μm with tolerances of ± 25 μm can be manufactured by this process [60].

4.2.4.6 Fused Deposition Modeling

FDM works on the principle of heating the material which is generally in the form of a thermoplastic filament or metallic wire to its melting point and then to extrude it layer by layer to manufacture a 3D part [49]. This ALM process was developed by S. Scott Crump (cofounder and chairman of Stratasys Inc.) in the late 1980s and was commercialized in 1990 by Stratasys Inc. It is also known as *Plastic Jet Printing (PJP)*. Fig. 4.20 presents a schematic of the typical FDM system layout and the working principle. A typical FDM machine consists of extrusion nozzles, spools of base and support materials, a build platform, and a control system. The FDM process commences with unwinding the plastic filament or metallic wire

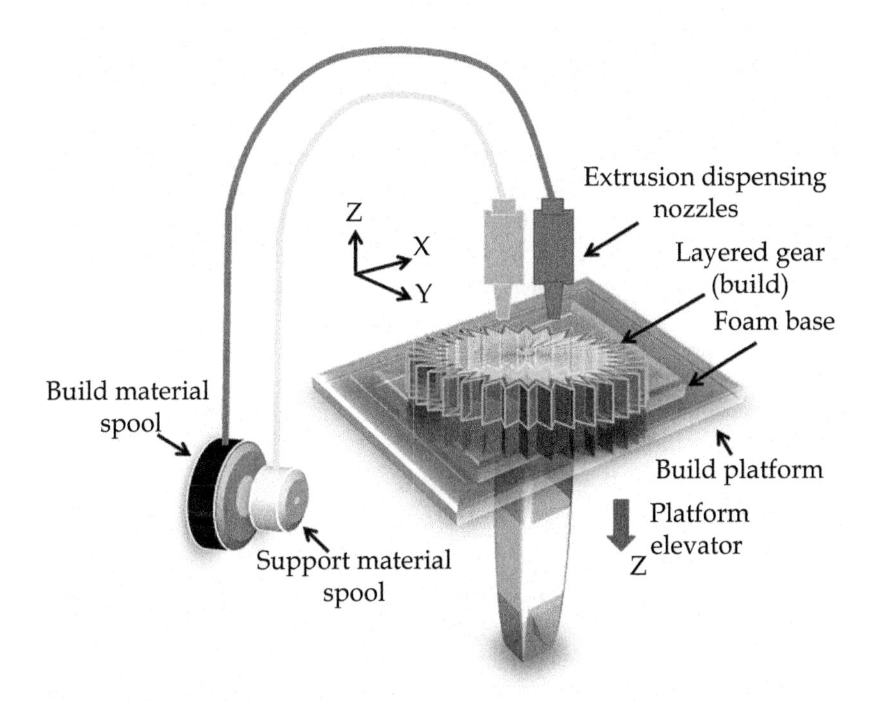

FIGURE 4.20 Gear manufacture by fused deposition modeling (FDM).

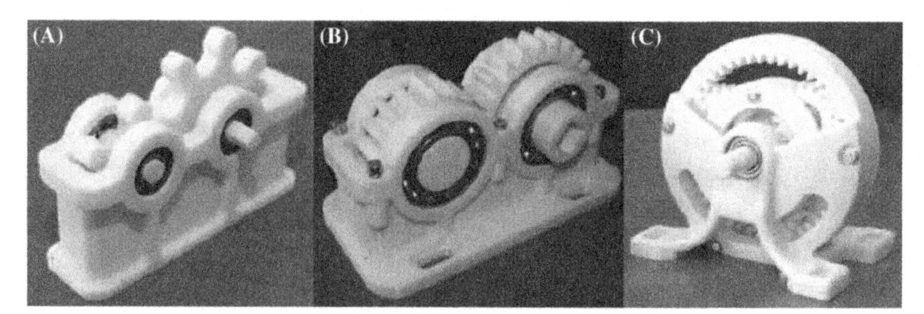

FIGURE 4.21 Gearing systems manufactured by FDM: (A) single-stage gear with cycloidal profile; (B) single-stage gear with involute profile; and (C) planetary gear. *From G. Budzik, The use of the rapid prototyping method for the manufacture and examination of gear wheels, in: M. Hoque (Ed.), Advanced Applications of Rapid Prototyping Technology in Modern Engineering, InTech, Croatia, 2011, ISBN: 978-953-307-698-0, © Budzik G published under CC BY 3.0 license, available from: http://dx.doi.org/10.5772/22848.*

from its spool and supply to the extrusion nozzle. The nozzle melts the material and extrudes it onto a build platform or table. The movements of both nozzles and the build platform are controlled to follow the CAD-based CNC tool path along the X, Y, and Z coordinates. Depending on the requirement a support material is also supplied via an adjacent nozzle. The nozzles are mounted in the extrusion head which also feeds the filament into the nozzle. The extrusion nozzles trace out an appropriate layer on the platform before it is vertically displaced in preparation for the next layer. The extruded layer of plastic/metal cools and hardens, immediately binding to the layer beneath it. The cycle is repeated and the layers are extruded until the entire specified geometry of the gear is manufactured [48,49]. Commercial thermoplastics such as ABS, polyamide, polycarbonate, and polyetherimide are frequently used to manufacture gears by FDM [48,59]. Budzik [48] has demonstrated its ability to successfully manufacture selected gears (Fig. 4.21).

4.2.4.7 Advantages of Additive Layer Manufacturing for Gear Manufacturing

ALM offer the following advantages when associated with gear manufacturing:

- ALM technology is cleaner, simpler-to-use, material-efficient and eco-friendly. ALM-based processes are typically faster because they eliminate numerous intermediate steps required by conventional manufacturing processes and use digital data as input;
- ALM processes can manufacture complex gear geometries and shapes which cannot be manufactured otherwise;

- These processes can manufacture gears from a wide range of materials including metals and alloys, polymers, composites, ceramics, and some other exotic materials;
- They are resource-efficient; does not require complicated tools, lubricants, and extra resources;
- This technology involves short process chains that does not necessarily require management of repair, maintenance, and safe disposal of waste materials;
- With proper control and optimization of layer thickness and process parameter the ALM processes can yield good quality surface finish and close tolerances.

4.2.4.8 Factors Influencing Part Accuracy of Additive Layer Manufacturing

Special care and appropriate process control is required for ALM due to the properties of most of the materials being anisotropic, temperature sensitive, and sensitive to the operating environment. SLA and SLS offer advantages pertaining to material isotropy due to their binding mechanism when compared to 3D inkjet printing (3DP) in which binding is done by an adhesive [62]. Suitable deposition orientation, robust support structures, and holding arrangements are essential for a uniform cross-section and accuracy. Inappropriate tessellation (piecewise approximation of CAD models using triangles) may lead to increase nonconformity between the actual surfaces and approximated triangles and should be addressed at an early stage [63]. Error in slicing or choosing improper layer thickness is another factor which significantly influences the accuracy and finish of the gear. An adaptive slicing scheme which segments the CAD model with improved accuracy and surface finish without losing important features is recommended [64]. During part deposition generally two types of errors are observed, i.e., *curing errors* and *control errors*. Curing errors are due to over or under curing with respect to the curing time line whereas control errors are caused due to the variation in layer thickness or scan position control. Adjustment of the chamber temperature and an appropriately generous power source is needed for proper curing. Inadequate care in the temperature control frequently induces curling. The calibration of the system becomes mandatory to minimize control errors. Shrinkage also causes dimensional inaccuracy and can be addressed by appropriate scaling in x, y, and z directions [63–65]. ALM offers the ability to economically manufacture customized gears directly from digital data. Gears from microsize to bigger prototypes from a wide range of materials range can easily be manufactured by ALM processes. This technology is also increasingly employed to manufacture gears of non-standard tooth profile that is generally difficult to manufacture by conventional tooling. Dedicated ALM process specific control and implementation is a requirement for improved quality, high accuracy, and defect-free gears.

4.3 DEFORMING PROCESSES

Manufacturing by deformation involve plastic deformation of the workpiece material usually contained in a close or open die and is associated with large forming forces and maintaining a constant volume. Most of these processes result in recrystallization, grain refinement, and grain reorientation. Consequently, formed products may have improved mechanical properties.

4.3.1 Hot Embossing

Hot embossing is widely used for manufacturing precise micro or nanofeatures across a small area of a substrate. Typical applications include thermoplastic at an elevated temperature and approximately 10 bar pressures [66]. This process was developed in the 1970s at RCA Laboratories in Princeton, New Jersey (United States) [67].

4.3.1.1 Working Principle

Hot embossing involves pressing the workpiece material into a master die or mold to replicate the desired mold geometry. The working principle of hot embossing is presented in Fig. 4.22 [67,68]:

1. *Positioning of the thermoplastic sheet*: A micromold corresponding to the gear to be manufactured, is positioned over the polymer substrate to obtain a stack which is placed on a lower heating plate containing cooling channels. Oil with a high heat capacity is circulated in cooling channels to ensure isothermal heating and cooling of the mold and the polymer substrate.
2. *Heating to molding temperature and subsequent molding:* Typically the lower heating plate is displaced vertically to heat up the stack while maintaining an appropriate pressure (pneumatically or hydraulically). To ensure effective reproduction of all the details of the mold to the substrate the mold and substrate are heated beyond the glass transition temperature of the substrate material. This softens the substrate material and minimizes the force required to fill all cavities of the mold.
3. *Cooling to demolding temperature*: The embossed or molded gear is cooled to the demolding temperature by the cooling plates maintaining the pressure constant.
4. *Demolding*: It involves removal of the hot embossed gear from the mold.

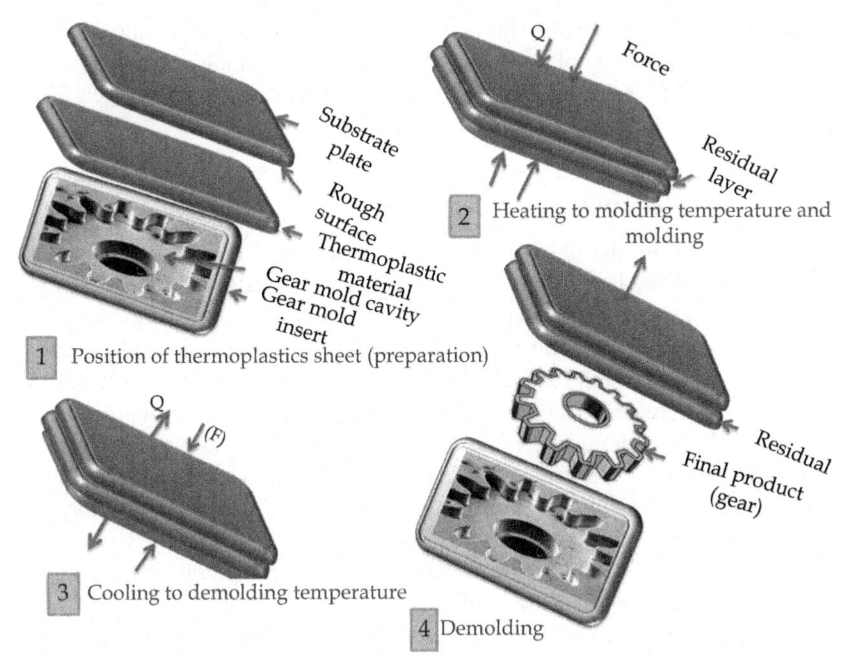

FIGURE 4.22 Schematic of working principle of hot embossing process. *From N.K. Jain, S.K. Chaubey, Review of miniature gear manufacturing, in: M.S.J. Hashmi (Ed.), Comprehensive Materials Finishing, vol. 1, Elsevier, Oxford, UK, 2016, pp. 504–538, doi:10.1016/B978-0-12-803581-8.09159-1. Elsevier © 2017. Reprinted with permission.*

4.3.1.2 Capabilities, Advantages, Limitations, and Applications

Hot embossing has the following distinct advantages:

- This process is the best and cost-effective reproduction or copying process for high aspect ratio microgears having higher dimensional accuracy;
- It can reproduce arrays of microfeatures including microgears on the substrates using a single master die;
- It is more suitable for high volume production due to simple setup and shorter lead time;
- It provides improved mold life because it can be used many times due to low wear;
- It is highly cost-effective for microgear production;

Hot embossing has the following limitations:

- Typically used for thermoplastics only;
- It is difficult to attain a sufficiently high embossing pressure;
- Tool and die making and the ancillary equipment associated with this technique are expensive and complex.

Hot embossing can be used for both amorphous and semicrystalline thermoplastics for a wide range of embossing temperatures. Examples are microgears as manufactured by Kuo and Zhuang [69]. It is a good choice for microgears for MEMS devices, biomedical, scientific, and electronic instruments [70,71].

4.3.2 Fine Blanking

Fine blanking is similar to extrusion where the metal is extruded through die cavities to form the desired gear tooth geometry. It is also similar to stamping but differs in the sense that it uses two opposing dies to form the gear from sheet material. The working principle fine blanking of a typical spur gear is presented in Fig. 4.23 [72,73]. Initially the gear blank, in the form of a sheet, is compressed between the blank holder and die. The impingement ring in the bottom die counters radial material inflow. The punch then displaces upward until the gear is fully sheared. The counter punch then descends, blanking pressure is reversed, punches retract, the gear is removed and the machine is readied for the next cycle.

This process is suitable to manufacture spur gears in high volumes. By making provision for an additional rotational movement of the punch and counter punch, this process can be extended for the manufacture of helical gears (rotational fine blanking) [72].

Fine blanking may be used to manufacture bevels, multiple, and complex gear sets which are difficult to produce by stamping. Gears made by fine

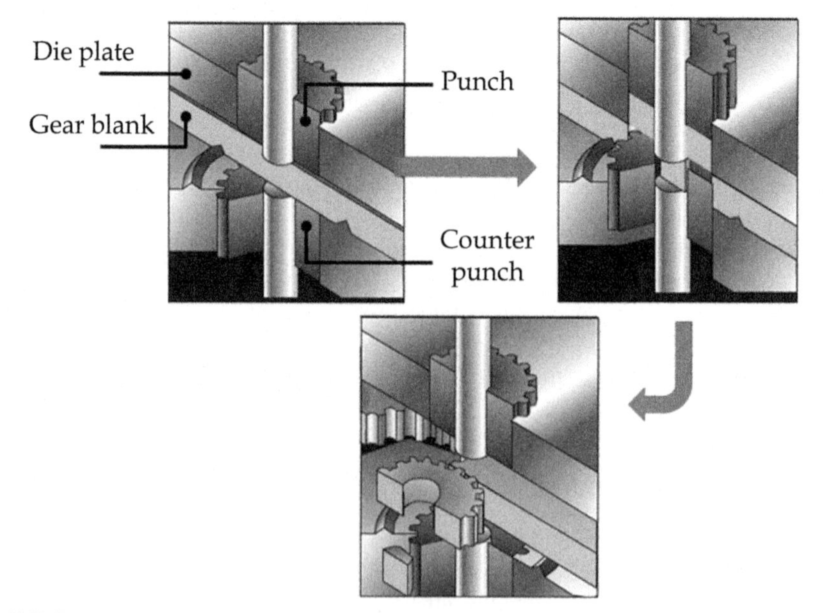

FIGURE 4.23 Working principle of fine blanking of gears.

FIGURE 4.24 Examples of fine-blanked gears.

blanking possess excellent dimensional accuracy and surface finish and have comparatively higher strength than those made by stamping and powder metallurgy. Fine-blanked gears have widespread applications including automotive, office equipment, appliances, medical equipment, electronic instruments, etc. [74]. Fig. 4.24 displays a variety of gears manufactured by fine blanking.

4.4 HYBRID PROCESSES

Certain advanced manufacturing processes have developed by combining a conventional process with an advanced process or combining two or more of the advanced processes.

LIGA is an abbreviation of the German words lithographie, galvanoformung and abformung. The English translation is lithography, electroplating, and molding. It is one of the most frequently used advanced processes for manufacturing microgears and micropart of high aspect ratio [75].

4.4.1 Lithographie, Galvanoformung and Abformung

Romankiw et al. (IBM in 1975) was the first to combine electroplating (deposition) and X-ray lithography [76]. The LIGA process in its present form was developed at the Karlsruhe Nuclear Research Centre (Germany) to manufacture nozzles for the uranium enrichment process in the 1990s [77].

4.4.1.1 Working Principle

The steps involved to manufacture a microgear by LIGA are shown in Fig. 4.25 [79,80].

1. *X-ray mask*: A mask with the complementary shape of the gear is prepared by using a thin film of gold on beryllium. This ensures that the X-ray mask allows transmission of the appropriate X-rays to the photoresist material only.
2. *Substrate preparation*: Deep X-ray lithography (DXL) is then used to process and develop the desired shape of the gear in an X-ray sensitive

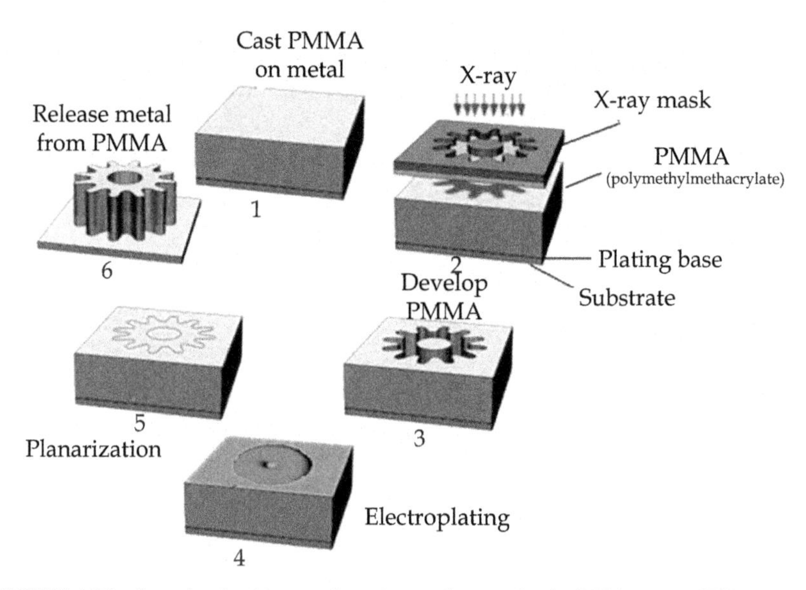

FIGURE 4.25 Steps involved in manufacturing a microgear by the LIGA process [78].

photoresist material. The desired properties of the photoresist material are (1) thermally stable up to elevated temperatures; (2) excellent adhesion properties during the electroplating process; (3) dry and wet etching resistant; and (4) insolubility of the unexposed "resist" material during the development process. Generally, polymethylmethacrylate, ployvinylidene fluoride, polycarbonate, polysulfones, epoxy phenol resin, plexiglass, and polyether ketones are used as photoresist material. The photoresist of the chosen material and the required size as per the dimensions of the gear is prepared by removing foreign material particles from its top surface and placing it on a base material which can either be a conducting or electrically conductive coated insulator. This is required to facilitate the subsequent electrodeposition in the prepared mold of photoresist. Titanium, nickel, austenitic steels, copper plated with gold and silicon wafers with thin titanium layers are used as base materials. Sometime, glass plates with thin metallic coating can also be used as base material. The combination of the base material and photoresist is referred to as the substrate.

3. *Transfer of the pattern by X-ray exposure of the substrate*: The substrate is exposed to parallel X-rays of short wavelength (provides for deep penetration) via the openings in the mask. Short wavelength X-rays are able to provide exposure accuracies of 0.2 μm and aspect ratios of a 100. This is essential for the high resolution exposure of high aspect ratio, sharp, deep, and thin cavities of the desired gear shape into the photoresist material.

4. *Development of the mold made of photoresist material*: The exposed or unexposed (depending on the type of photoresist material) photoresist material is then chemically removed to expose the desired gear geometry in the form of an accurate cavity (mold) that contains the outline of the gear.
5. *Electrodeposition of the gear/part material in the mold*: The photoresist mold is then "filled" in by an electrodeposition process where it is suspended in an electrodeposition bath in which the gear material is deposited from the bottom-up.
6. *Planarization*: Removal of the excess electrolytically deposited material from the top surface of the mold to obtain a flat surface.
7. *Releasing the final product*: Stripping of the surrounding photoresist material from the manufactured gear.

4.4.1.2 Capabilities, Advantages, Limitations, and Applications

The LIGA process offers the following capabilities and advantages:

- This process is able to manufacture high precision and high aspect ratio microgears made of metallic materials and plastics with dimensional accuracies of less than 1 μm and nanometer scale surface finishes (average surface roughness value <50 nm);
- It offers enhanced flexibility in geometry;
- Does not introduce any thermal distortion;
- It is suitable for mass production;
- The following are the limitations of the LIGA process;
- LIGA requires high energy X-rays which poses a radiation threat to the operator that must be addressed;
- It is a complicated multistep process;
- It is challenging to obtain very close tolerances for microgears;
- LIGA process equipment is expensive.

LIGA can be used to manufacture microgears with complex shapes and geometries for many scientific and industrial applications.

4.5 SUSTAINABLE MANUFACTURING OF GEARS

Environment awareness is of the utmost importance to all socially responsible manufacturers. To be competitive on a global scale the manufacturing needs to be aligned with various strict environmental regulations. The gear manufacturing industry at large along with downstream gear users in other industrial and manufacturing segments at large is striving to improve productivity and product quality while simultaneously maintaining a clean and sustainable environment. This can only be achieved by adopting sustainable techniques of manufacturing which include the use of environment-friendly lubricants and lubrication techniques such as dry-cutting (DC), minimum quantity lubrication (MQL) and cryogenic cooling while using improved tool materials and coatings in advanced and hybrid manufacturing processes [17].

4.5.1 Challenges and Opportunities

Subtractive or material removal type manufacturing processes constitute a major percentage of gear manufacturing with hobbing, milling, and shaping being the most extensively used gear machining processes. Fig. 4.26 depicts the typical processing sequence to manufacture a gear by the machining process. It commences with preparation of the gear blank from the selected gear material, followed by cutting of the gear teeth, followed by heat treatment, if required, and finishing processes to impart the desired mechanical properties (both surface and interior). The manufacturing needs to produce a gear with the desired quality, i.e., acceptable mechanical properties, appropriate dimensional tolerances, and surface integrity including roughness and other descriptors including residual stress.

The requirements of the gear finishing have a significant impact on the overall sustainability because it usually implies increased tool wear, higher consumption of the cutting fluid and energy and excessive waste management including handling, disposal, and recycling [17]. The problems associated with manufacturing of gears by conventional processes can be summarized as follows:

- All conventional processes for manufacturing of gears necessitate finishing processes such as trimming, shaving, grinding, honing, burnishing, etc. to obtain the desired gear quality.
- Gear finishing processes require fabrication, repair, and maintenance of the finishing tool, consume large amounts of cutting fluid and energy, increases the burden of handling, recycling and disposal of waste, and escalates the overall cost.
- Poor lubrication ability of conventional lubricants and lubrication/cooling systems imparts excessive tool wear during machining and affects the sustainability by necessitating resharpening facilities and/or new tools.

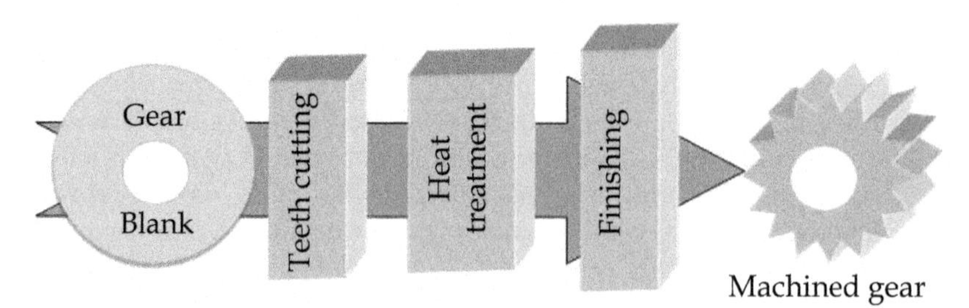

FIGURE 4.26 Typical sequence of processes to manufacture a gear by machining [17].

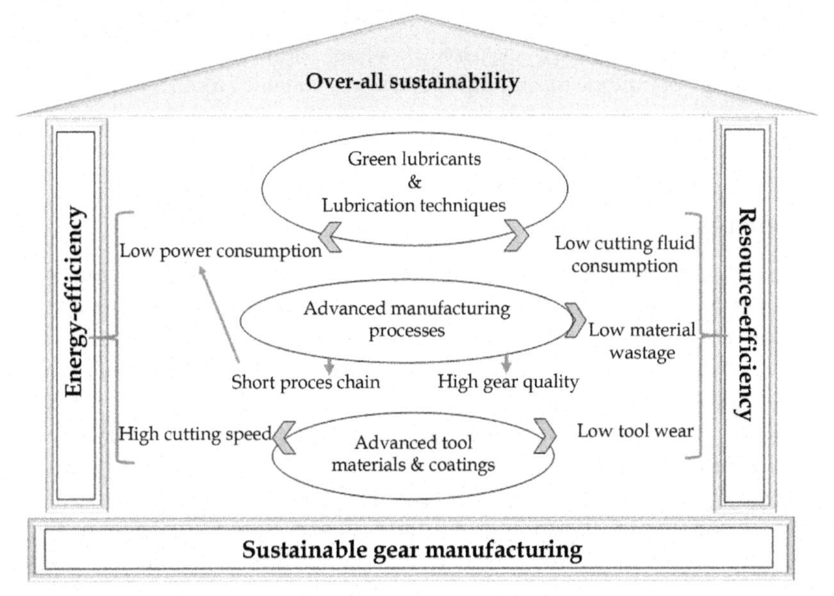

FIGURE 4.27 House of sustainability representing manufacturing strategies to achieve overall sustainability in gear manufacturing [17].

- Supplying and pumping significant quantities of cutting fluid to the cutting zone during conventional lubrication and cooling implies elevated power consumption and may also affect the health and safety of the operator if the fluid has toxic properties.
- Handling and disposal of wet chips (metal cutting wastes) and used cutting fluid are difficult tasks to deal with and may consequently affect the environment.

These factors all adversely affect the overall sustainability in gear production.

Considering the aforementioned challenges, engineers, and technologists are developing sustainable manufacturing strategies to (1) shorten the long process chain of gear manufacturing; (2) minimize the use and consumption of harmful cutting fluids; (3) reduce the chances of tool failure; and (4) produce tolerances close to the net shaped or near-net shaped gears. Fig. 4.27 presents selected important manufacturing strategies and their consequences to achieve overall sustainability in gear manufacturing. These include

1. Adopting alternative advanced and hybrid processes of gear manufacturing such as gear rolling, and WSEM, etc.;
2. Minimizing use and quantity of the harmful cutting fluids by using environment-friendly lubricants such as vegetable oils, synthetic esters,

and other biodegradable lubricants and employing sustainable lubrication techniques such as MQL, cryogenic cooling, and DC;

3. Selecting optimum machining condition, suitable tool materials and coatings.

4.5.2 Environment-Friendly Lubricants and Lubrication Techniques

As discussed in the forgoing sections, the use of large amounts of mineral-based cutting fluids may adversely impact the environment because it may lead to increased ground contamination, increased energy consumption, increased wet-chip handling and waste disposal, and increased health and safety risk [81,82]. To address these issues, there has been a steadily increasing quest to perform machining operations in dry or near-dry condition. During dry machining, no cutting fluid is used for any cooling and/or lubrication whereas, for near-dry machining "green" cutting fluids such as vegetable oils, synthetic esters, and fatty alcohols are used because they are more environmentally-friendly, biodegradable, and therefore less harmful to the human beings, plants, and animals instead of synthetic (mineral oil based) cutting fluids [81,82]. Vegetable oils generally possess excellent lubricant properties such as a high flash point to reduce smoke formation and fire hazards, high viscosity index which provides more stable lubricity over the operating temperature range, high boiling point, and molecular weight which results in less loss from vaporization and misting, excellent solvency for lubricant additives and easier miscibility with other fluids. Vegetable oils, especially those derived from rapeseed, soyabean, castor, and palm oils are some of the more promising candidates as biodegradable lubricants. Synthetic esters also have very good biodegradability as well as low toxicity, which make them desirable as a green lubricant. They are synthesized by the chemical reaction of an alcohol and a fatty acid which can take place either thermally or with the use of a catalyst.

With regards to gear manufacturing, environmentally-friendly machining implies DC of gears, i.e., dry-hobbing, dry-shaping, dry-milling; near-dry or MQL assisted; cryogenically cooled and combined MQL-cryogenic assisted gear machining.

4.5.2.1 Dry-Hobbing

4.5.2.1.1 Overview

Considering the challenges faced in conventional gear hobbing and to overcome its limitations, the concept of dry-hobbing of gears was introduced by Liebherr Verzahntechnik GmbH in 1993 [83]. Significant research and development work in this field has been contributed by Mitsubishi Heavy Industries Ltd. which in 1997 introduced the world's first high production rate dry-hobbing system for manufacturing gears.

Initially, carbide cutters were used in a dry-hobbing process in which the cutting tool must withstand the thermal and mechanical stresses of the process. Considering the problems associated with a carbide hob cutter that includes higher cost and unexpected chipping; titanium nitride (TiN), titanium cyanide (TiCN), and titanium aluminide (TiAlN) coated high-speed steel and carbide hob cutters were developed in order to ensure a sustainable tool life. These coated hobbing cutters had excellent heat and wear resistance due to coating and demonstrated a substantial increase in the tool life when used to machine gears at cutting speeds much higher than that of conventional hobbing [84]. TiAlN is the most preferentially used coating which offers high hardness levels of 3300–3500 HV complemented by excellent sliding characteristics and operating temperatures up to 900°C [83]. In one documented work, Mitsubishi Heavy Industries Ltd is reported to have achieved a reduction of total hobbing cost by 34%, a twofold increase in cutting speed, a five times tool life increase and 51% reduction in electricity cost after utilizing a dry-hobbing machine tool equipped with a TiAlN-coated hobbing cutter made of high-speed steel [84,85]. Dry hobbing is essentially a hobbing process where hardened or unhardened gear blanks (depending upon requirements) are machined with coated cutters/hobs in the absence of any cutting oil. The dry hobbing machines operate at cutting speeds between 90 and 250 m/min and in some cases even higher.

4.5.2.1.2 Requirements for Effective Dry-Hobbing

Thermal stability, accuracy, and fast and efficient chip removal are the three major design considerations for an effective dry-hobbing machine tool (see Fig. 4.28). The design of the dry-hobbing machine tool must be such that thermal stability is ensured because there is no coolant in the machine bed to

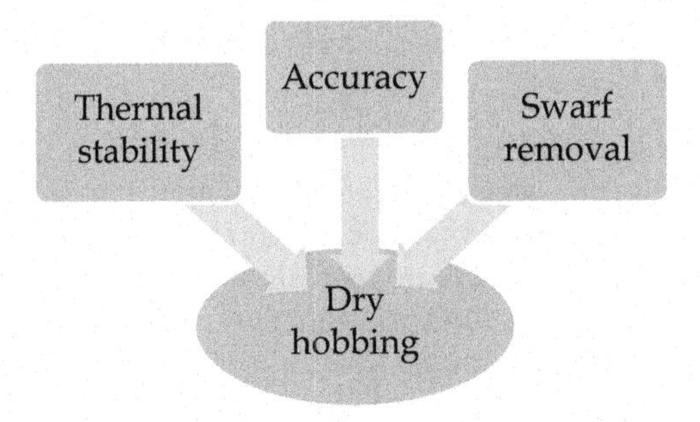

FIGURE 4.28 Essential design consideratios for a dry-hobbing machine tool.

remove heat generated by the cutting process. These machines are designed with special internal ribbing for thermal stability. They are generally quite heavy and incorporate special steep walls around the tables. They also include an integrated chip conveyor to not only remove the chips for disposal but also to effectively control the heat load associated with the chips. In some cases, special vibration panels underneath the workpiece are used to accelerate chip removal. Most of the heat generated in the dry-hobbing process (as much as 80%) is carried away by the chips [86]. These machines are operated at ultra-high speeds; therefore they need to be accurate enough (in terms of machine kinematics, rigidity, and work holding fixtures, etc.) to achieve exacting tolerances.

From safety point of view, appropriate swarf (waste chips) removal and protection against its ill effects (damage to man and machine) are required. It involves sealing the cutting area, fitment of air-blowing nozzles, and integrated chip conveyors to remove waste chips, employing safety covers, guards, and dust collectors at suitable locations in the machine tool, etc.

4.5.2.1.3 Benefits

Dry-hobbing not only dramatically reduces environmental risks by eliminating the need for cutting fluids and wet-chip disposal, but also effectively improves the machining efficiency and reduces the manufacturing cost by ensuring operation at higher cutting speeds and with longer tool life. The following points highlight the tangible benefits of dry-hobbing:

- Improved cutting efficiency;
- Improved gear quality;
- Potential production cost reduction as a result of reduced manufacturing cycle time, extended tool life, eliminating pumping and circulatory system for the cutting fluid;
- Total elimination of cutting fluid management including their disposal and recycling;
- Environmentally-friendly and safer for the operator. The operator can work in a cleaner environment; moreover, shop floors and equipment are not covered with coolant.

With increased use of the dry-hobbing, dry-cutting conditions have also been developed and subsequently employed for other gear manufacturing processes and machine tools, i.e., gear shaping, and bevel and hypoid gear generation, etc. In general, dry-gear cutting has presented a major breakthrough in productivity, economy, and ecology (Fig. 4.29). It is an environment-friendly alternative and offers many advantages and very few disadvantages over conventional wet cutting/machining of the gears.

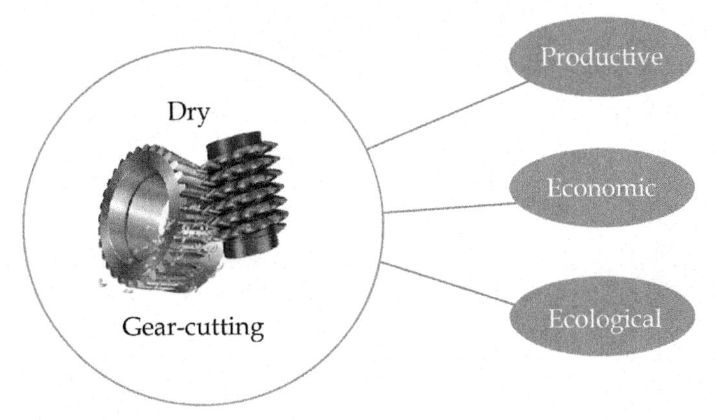

FIGURE 4.29 Benefits of dry-cutting of gears [17].

4.5.2.2 Minimum Quantity Lubrication Assisted Machining of Gears

4.5.2.2.1 Overview

MQL is a microlubrication technique that facilitates near-dry machining. It eliminates large quantities of water and mineral oil-based cutting fluids and replaces them with a small quantity of environment-friendly lubricant (mostly vegetable oils) mixed with air. In MQL-assisted machining, a small amount (10–250 mL/h) of cutting fluid is introduced to the chip-tool interface region along with compressed air, which acts as the carrier medium, as a replacement for several liters per hour of conventional coolant. Supplying a small amount of cutting fluid is less expensive and consumes less power and thus reduces the cost. The main benefit of MQL is that it primarily focuses on improving the frictional behavior therefore controlling the heat generation at its source rather than just trying to remove as much heat as possible such as conventional cooling does. This results in improved tool life and good workpiece surface integrity. The chips generated during MQL machining are nearly dry and are easy to recycle. The cutting performance and the overall quality of the parts manufactured by utilizing MQL therefore depends on optimizing the appropriate process parameters including the type and flow rate of the lubricant and the nozzle position and pressure [87–90]. The multiperformance ability of MQL machining such as heat management, cutting interface lubrication, environment-friendliness, and energy efficiency are the primary motivation for its application in gear machining [81,82].

4.5.2.2.2 Working Principle and Mechanism

Fig. 4.30 depicts different components of a typical MQL system which includes an oil reservoir, source of compressed air, MQL device, piping, and nozzle. Effective lubrication between the tool and workpiece gear is achieved

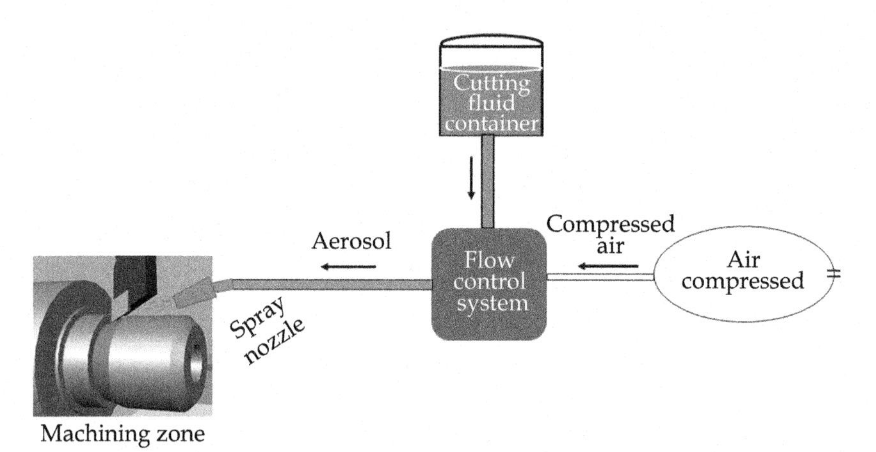

FIGURE 4.30 Schematic representation of various components of an MQL system [17].

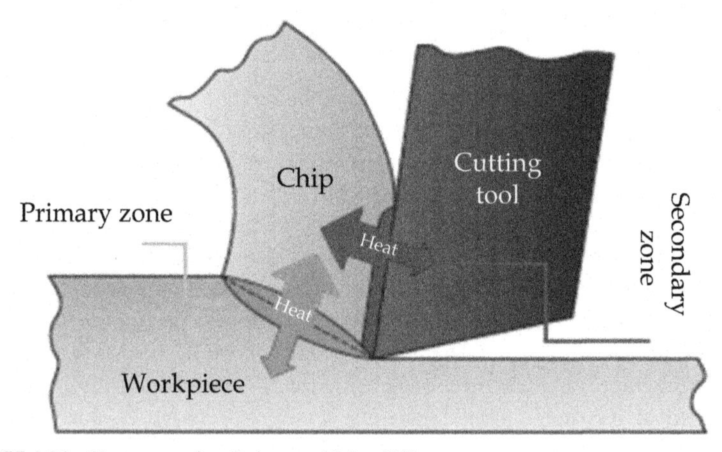

FIGURE 4.31 Heat generation during machining [17].

with a flow of compressed air containing finely dispersed droplets of oil which is known as aerosol. The MQL device consists of a micropump and nozzles and a special aerosol generator system that can produce aerosol with an oil droplet size of about 0.5–5 μm by mixing lubricant oil and compressed air. Owing to this small size the droplets of oil have hardly any inertia or rate of fall and are transported by compressed air to the zones of heat generation in machining process, i.e., tool–chip and tool-workpiece interfaces.

Fig. 4.31 depicts the two most important shear zones that are the main sources of heat generation during a conventional cutting process. The heat generated at the primary cutting zone due to shearing (atomic slip by largely dislocation movement) of the workpiece material is unavoidable. The generation and transfer of heat at the secondary shear zone, i.e., tool-chip interface

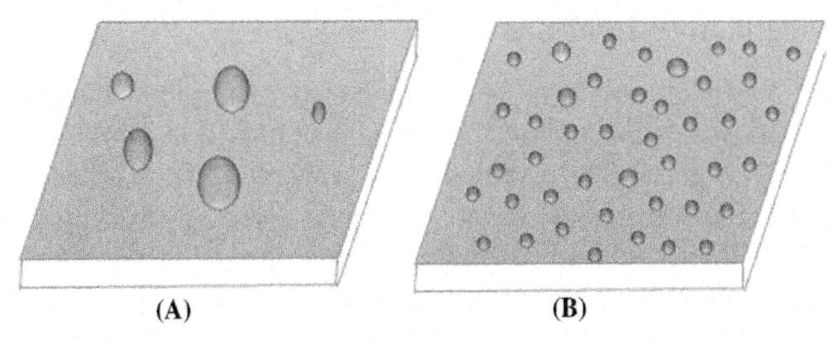

FIGURE 4.32 (A) Poor wetting of the workpiece and tool due to uncontrolled atomization of the air/oil droplets at the nozzle in conventional cooling; (B) MQL droplets wet the workpiece evenly due to much smaller, homogenous droplets [17].

can however be managed with improved lubrication. Fundamentally, the small size and distribution of the oil droplets in the MQL aerosol are homogenous (since the aerosol is atomized under controlled conditions) and therefore results in a high degree of surface wetting as well as the ability to reach to inaccessible regions on the workpiece (Fig. 4.32B). The friction and thus generation and transfer of heat from the chip to the tool and workpiece gear, is reduced. Optimal lubrication during removal of the chips in the chip groove not only permits higher machining speeds but also results in an improved gear surface finish. In conventional cooling/lubrication, the coolant/lubricant particles are comparatively big and massive therefore, uniform distribution over the surfaces of tool and gear becomes problematic and only partial lubrication occurs (Fig. 4.32A) which is ineffective to minimize heat generation.

Another lubrication strategy, i.e., minimum quantity cooling and lubrication is used to supply a mixture of cryogenically cooled low temperature (0–60°C) air and vegetable oil in a small quantity to the machining zone. Generally, cryogenic cooling is done with liquid nitrogen. This microhybrid lubrication technique is adopted to address both cooling as well as lubrication simultaneously while machining.

The mixture of cryogenic air and vegetable oil is sprayed in the form of microdroplets into the machining zone where it absorbs the heat generated during machining as a coolant, and becoming a form of gaseous-fluid protection layer between the chip and the tool face that functions as a lubricant; this lubricant may further reduce the tool–chip interface temperature and thus chemical reaction between tool and chip. This reduces the adhesion and diffusion of the cutting tool and hence increases the tool life [81,82].

These lubrication techniques may achieve effective cooling and lubrication of the gear machining processes even with small quantities of oil allowing machining to be conducted at higher cutting speeds and achieving longer tool lives without the necessity to condition or dispose-off cooling lubricants thus ensuring higher productivity, better product quality, and environment-friendliness.

4.5.3 Recent Investigations

A series of investigations have been conducted on implementation of dry-cutting, MQL-assisted machining, cryogenic cooling, and combined cooling and lubrication of gears. These include analyzing the effects of environment-friendly lubrication techniques on tool wear and surface finish, thermal stress generation, and thermo-elastic displacements on geometric accuracy of gears [87–89]; effect of MQL-assisted machining on wear of the hobbing cutter and cutting forces during hobbing [91]; selecting the optimal parameters for gear hobbing under cryogenically coupled MQL cutting of gears [92]; and comparative study between wet gear hobbing and cryogenic gear hobbing with MQL [93]. The following paragraphs summarize these recent investigations.

Fratila [87] conducted experiments to investigate the effects of dry and near-dry machining processes on gear milling efficiency. The experiments performed were focused on the technical evaluation of cooling and lubrication effects, cutting force behavior and tool wear, while also emphasizing the effects of the above-mentioned parameters on process quality by means of surface quality evaluation. He monitored the emitted particulates and gases in the environment during the process. A total of 400 right hand helical gears (material: 16MnCr5; module: 2.75 mm; outside diameter: 110 mm; number of teeth: 37, helix angle: $18°50'$) were machined on a CNC gear milling machine by using five cooling and lubrication techniques namely flood lubrication (FL), DC, MQL, minimal cooling technique (MCT), and minimal quantity cooling and lubrication (MQCL). Eighty gears were machined corresponding to each technique. The following cutting fluid media were used: FL, rotanor oil (fluid flow—100 l/min); MQL, vegetable oil (fluid flow—0.4 mL/min); MCT, emulsion with anticorrosion additives (fluid flow—25 mL/min) and MQCL, vegetable oil, and compressed air (5 bar). The outstanding lubricating action of the MQL system resulted in minimum wear of the hobbing cutter along with satisfactory surface roughness of the gear teeth. FL- and DC-produced satisfactory surface finish at the gear flanks but resulted in higher wear of the hobbing cutter. An analysis of the cooling effect by measuring the gear temperature found a significant temperature rise for MQL, exhibiting its reduced cooling effect and its close similarity with DC. FL, MQCL, and MCT were all found to have nearly similar temperature increases. The level of environment pollution corresponding to all the lubrication techniques was monitored by air volume analysis through a multigas monitor device. The concentration of pollutant gases and particulates demonstrated no significant pollution effects of MQL equipped gear milling when compared with the other techniques. As a final conclusion, it was reported by Fartila [87] that MQL equipped gear milling is a good alternative to the conventional (flooded) gear milling processes as it offers reduction of cost associated with lubricants and power consumption, increased tool life, improved surface quality of gears, improvements in the chip handling and recycling process and a cleaner environment which is the most important.

Fratila and Radu [88] evaluated the extent that these environment-friendly lubrication techniques can solve the requirements of the cutting fluids, i.e., cooling, lubrication, and what are the consequences of thermal stress and thermo-elastic displacements during the gear milling process. Finite element analysis followed by a series of experiments was conducted for the thermal investigations. The experimental work performed employed a lubrication device equipped with two nozzles for MQL whereas two devices were used with additional pressure for MCT. Raps oil, a green lubricant, a kind of vegetable oil (0.4 mL/min) was used as a lubricant during MQL. Water (20 mL/min) was added as an auxiliary lubricant in MCT. The cooling and lubrication effects were then evaluated separately for MQL and MCT and together in MQCL for comparison purpose. Gear temperature was measured with a contact thermometer immediately after the completion of the milling operation. Steady state finite element thermal models corresponding to all lubrication-cooling techniques, viz., MQL, MCT, MQCL, FL, and DC were developed using the experimental results. Theoretical analysis of thermo-elastic strains using the finite element method confirmed the results of experimental research. The conclusions of thermal study confirmed the results obtained in the experimental research, i.e., cooling and lubrication are unsatisfactory at DC and the combination of minimal cooling and lubricating improved both the cutting conditions and the process accuracy. In terms of heat stress in the wheel at the end of gear milling the cutting with flood cooling was found to be the most unfavorable case. Intense cooling caused a high-temperature gradient which led to increased thermal strain. The lowest values of thermal stress were obtained for DC but the process was adverse with reference to thermo-elastic displacements. It was summarized that the MQL and MQCL were favorable alternatives and the strains and thermal stresses displayed intermediate values in comparison with DC and FL-based gear milling.

Another investigation by Fratila [89] focussed on finding the effects of environment-friendly lubrication techniques on the geometric accuracy of milled gears. The manufacturing quality of any gear is mainly affected by the cutting tool condition, tool and gear blank spindle position and induced vibrations in the machine tool [32]. A similar set of parameters and experimental conditions as reported by Fratila [87] were employed while investigating the effects of cooling and lubrication techniques on gear accuracy. Tooth thickness, span, total tooth profile deviation, total tooth flank line deviation, and pitch deviation were considered as the parameters of gear geometric accuracy. Results of the investigations revealed that for all variants of cooling/lubrication processes, the gear quality in terms of tooth profile, total tooth flank line and pitch deviation was similar to that obtained in the FL processing. However, the variation analysis of tooth distances on gear circumference identified the superiority of MQL over conventional cooling methods. In terms of tooth thickness the performance of flood cooling was the best, followed by MCQL and MCT. Finally, it was concluded that only one parameter of gear accuracy i.e. tooth thickness was affected by the lubrication techniques and that gear milling with a DC environment was not recommended for good quality of the milled gears.

Stachurski [91] investigated the effects of MQL on hobbing cutter wear and cutting force during gear hobbing. Spur gears made of two different materials, viz., normalized carbon steel and alloy steel 42CrMo4 were hobbed with a high speed steel hobbing cutter under the influence of MQL and conventional flood cooling. The flow rate of oil during MQL was 25 mL/h and in flood cooling was 10 L/min. Similar hob wear rates were obtained for MQL and conventional flood cooling. Progression of cutting forces with hobbing time for both methods was comparable and on that basis alone application of the MQL method in gear hobbing was recommended as an economic and environment-friendly alternative.

A study reported by Zhang and Wei [92] on cooled air MQL gear hobbing highlights some important aspects of simultaneous minimization of gear teeth surface roughness and hobbing cutter wear. A Taguchi technique integrated utility concept was employed for multiresponse optimization of the gear hobbing process. Their study also aimed to determine the optimum amount of MQL, the most appropriate cold air temperature and feed rate during hobbing of medium carbon steel (C45) by YG6X hard alloy tool. The results of experimental investigations and optimization revealed 40 mL/h MQL flow rate, −45°C cold air temperature and 0.2 mm/rev feed rate as the optimum parameters to simultaneously minimize wear of the hobbing cutter and surface roughness of gear teeth.

There have been some recent investigations not specifically based on gear manufacturing but rather from the perspective of complete powertrain machining, e.g., crankshaft manufacturing, machining of aluminum components and fabrication of gear-box housings by MQL techniques [90,94] which also advocates the conversion of whole powertrain machining from conventional to MQL-based machining. Other significant developments on efficient and sustainable gear manufacturing systems are ongoing as part of research and development of various large gear manufacturing industries but due to the proprietary nature of the technologies and professional competitiveness those are not available or disclosed in the public domain.

REFERENCES

[1] J. Hecht, Short history of laser development, Opt. Eng. 49 (9) (2010) 091002 (doi:10.1117/1.3483597).

[2] http://www.photonics.com/Article.aspx?AID = 42279.

[3] http://aml.engineering.columbia.edu/ntm/level1/ch02/html/l1c02s01.html.

[4] C. Cooper, Gear fundamentals: alternative gear manufacturing, Gear Technology Magazine, November/December (1998) 9–16.

[5] J. Lawrence, Advances in lasers materials processing-technology, Research and Application, Woodhead Publishing, Cambridge, 2010. ISBN: 9781845694746.

[6] G. Chryssolouris, Laser Machining: Theory and Practice, Springer-Verlag, New York, 1991. ISBN: 978-1-4757-4086-8.

[7] https://www.epiloglaser.com/products/co2-laser-systems.htm.

[8] R. Schaeffer, Fundamentals of Laser Micromachining, CRC Press, Boca Raton, FL, 2012. ISBN: 9781439860557.

[9] U. Klotzbach, A.F. Lasagni, M. Panzner, V. Franke, Laser micromachining, in: F.A. Lasagni, A.F. Lasagni (Eds.), Fabrication and Characterization in the Micro-Nano Range, Springer-Verlag, Berlin, Heidelberg, 2011, pp. 29–46.

[10] H. Liess, J. Heinzl Manufacturing of micro-ridges and micro-gears by laser ablation, in: Proceedings of ASME 7th biennial conference on engineering systems design and analysis, Manchester, England, 2004, July 19–22, 3: 305–309.

[11] P. Yang, G.R. Burns, J.A. Palmer, M.F. Harris, K.L. McDaniel, J. Guo, et al., Microfabrication with femtosecond laser processing (A) laser ablation of ferrous alloys, (B) direct-write embedded optical waveguides and integrated optics in bulk glass. Sandia National Laboratories Report-5625, 2004.

[12] C.G. Khan, Malek, W. Pfleging, S. Roth, Laser micromachining of polymers, in: A.D. Campo, E. Arz (Eds.), Generating Micro- and Nanopatterns on Polymeric Materials, Wiley, Hoboken, NJ, 2010.

[13] C. Olsen, Manufacturing gears with waterJet machining, gear production supplement, Modern Machine Shop Magazine (2015) 16–24.

[14] H.T. Liu, E. Schubert, in: K. Mojtaba (Ed.), Micro Abrasive-Waterjet Technology, Micromachining Techniques for Fabrication of Micro and Nano Structures, InTech, 2012. ISBN:978-953-307-906-6

[15] M. Hashish, A model for abrasive water jet (AWJ) machining, Trans. ASME: J. Eng. Mater. Technol. 111 (2) (1989) 154–162.

[16] C. Birtu, V. Avramescu, Abrasive water jet cutting—technique, equipment, performances, Non-Convent. Technol. Rev. 16 (1) (2012) 40–46.

[17] K. Gupta, R.F. Laubscher, J.P. Davim, N.K. Jain, Recent developments in sustainable manufacturing of gears: a review, J. Cleaner Prod. 112 (4) (2016) 3320–3330.

[18] J.A. McGeough, Advanced Methods of Machining, Chapman and Hall Ltd, London, 1988.

[19] V.K. Jain, Advanced Machining Processes, Allied Publishers, New Delhi, 2002.

[20] K.H. Ho, S.T. Newman, S. Rahimifard, R.D. Allen, Start of the art in wire electrical discharge machining, Int. J. Mach. Tools Manuf. 44 (12-13) (2004) 1247–1259.

[21] K.H. Ho, S.T. Newman, State of the art electrical discharge machining (EDM), Int. J. Mach. Tools Manuf. 43 (13) (2003) 1287–1300.

[22] K. Gupta, Sujeet K Chaubey, N.K. Jain (2014), "Exploring wire-edm for manufacturing the high quality meso-gears" published in Procedia Materials Science (Elsevier) 5 (2014), 1755–1760. In Proceedings of International Conference on Advances in Manufacturing and Materials Engineering (ICAMME 2014), March 27–29, 2014, Surathkal, India.

[23] K. Gupta, N.K. Jain (2013), Deviations in geometry of miniature gears fabricated by wire electrical discharge machining. In Proceedings of International Mechanical Engineering Congress & Exposition (IMECE 2013) of ASME, V010T11A047, November 13–21, 2013, San Diego, California, USA.

[24] K. Gupta, N.K. Jain, R.F. Laubscher, Spark-erosion machining of miniature gears: a critical review, Int. J. Adv. Manuf. Technol. 80 (9-12) (2015) 1863–1877.

[25] K. Hori, Y. Murata, Wire electrical discharge machining of micro-involute gears, Trans. Jpn. Soc. Mech. Eng. Ser. C 60 (579) (1994) 3957–3962.

[26] K. Suzumori, K. Hori, Micro electrostatic wobble motor with toothed electrodes, in: Proceedings of 10th IEEE International Workshop on Micro Electro Mechanical Systems, Nagoya, Japan, 1997, pp. 227–232.

[27] H. Takeuchi, K. Nakamura, N. Shimizu, N. Shibaike, Optimization of mechanical interface for a practical micro-reducer, in: Proceedings of 13th IEEE International Conference on Micro Electro Mechanical Systems, Miyazaki, Japan, 2000.

[28] G.L. Benavides, L.F. Bieg, M.P. Saavedra, E.A. Bryce, High aspect ratio meso-scale parts enabled by wire micro-EDM, Microsystems Technology 8 (2002) 395–401.

[29] K. Gupta, N.K. Jain, Near Net Shape Manufacturing of Miniature Gears by Wire Spark Erosion Machining, Springer International Publishing AG, Switzerland, 2016 (ISBN: 978-3-319-25920-8 (Print); 978-3-319-25922-2 (Online)] (doi:10.1007/978-3-319-25922-2).

[30] K. Gupta, N.K. Jain, Analysis and optimization of the micro-geometry of miniature gears manufactured by wire electric discharge machining, Precis. Eng. 38 (4) (2014) 728–737.

[31] K. Gupta, N.K. Jain, Analysis and optimization of the surface finish of wire electrical discharge machined miniature gears, Proc. IMechE: B J. Eng. Manuf. 228 (5) (2014) 673–681.

[32] K. Gupta, N.K. Jain, Comparative study of wire-EDM and hobbing for manufacturing high quality miniature gears, Mater. Manuf. Process. 29 (2014) 1470–1476.

[33] http://www.gfms.com/content/dam/gfac/proddb/edm/wire-cut/en/agiecharmilles-cut-1000-1000-oiltech_en.pdf.

[34] https://www.makino.com/about/news/Precision-Micro-Machining-Capabilities-Expand-with-Makino%E2%80%99s-UPN-01-Wire-EDM/586/.

[35] https://www.makino.com/about/news/trends-in-micro-machining-technologies/315/.

[36] N.K. Jain, S.K. Chaubey, Review of miniature gear manufacturing, in: M.S.J. Hashmi (Ed.), Comprehensive Materials Finishing, vol. 1, Elsevier, Oxford, UK, 2016, pp. 504–538 (doi:10.1016/B978-0-12-803581-8.09159-1).

[37] R. Tandon, Metal injection molding, Encyclopaed. Mater. Sci. Technol. (2001) 5439–5442.

[38] P. Divya, A. Singhal, D.K. Pattanayak, T.R.R. Mohan, Injection molding of titanium metal and AW-PMMA composite powders, Trends Biomater. Arti. Org. 18 (2) (2005) 247–253.

[39] P. Imgrund, A. Rota, L. Kramer, Processing and properties of bi-material parts by micro-metal injection moulding, in: Proceedings of 1st International Conference on Multi-material Micro Manufacture, 29th June–1st July, 2005, pp. 131–134, Forschungszentrum Karlsruhe, Karlsruhe, Germany.

[40] European Powder Metallurgy Association, Metal Injection Moulding: A Manufacturing Process for Precision Engineering Components, third ed., EPMA, Shrewsbury, 2013.

[41] B.A. Davis, R.P. Theriault, T.M. Osswald, Optimization of compression (injection/compression) molding process using numerical simulation, in ASME Conference, 1997.

[42] A. Shojaei, A numerical study of filling process through multilayer performs in resin injection/compression molding, Compos. Sci. Technol. 66 (2006) 1546–1557.

[43] V. Piotter, T. Gietzelt, L. Merz, Micro powder injection moulding of metals and ceramics, Sadhana 28 (1-2) (2003) 299–306.

[44] U.M. Attia, J.R. Alcock, A review of micro-powder injection moulding as a microfabrication technique, J. Micromech. Microeng. 21 (4) (2011) 1–41.

[45] J.G. Gutierrez, G.B. Stringari, I. Emri, Powder injection molding of metal and ceramics parts, in: J. Wang (Ed.), Some Critical Issues for Injection Molding, vol. 3, InTech, Croatia, 2012, pp. 65–88.

[46] R. Zauner, Micro powder injection moulding, Microelectron. Eng. 83 (4–9) (2006) 1442–1444.

[47] K. Nishiyabu, Micro metal powder injection molding, in: J. Wang (Ed.), Some Critical Issues for Injection Molding, vol. 5, InTech, Croatia, 2012, pp. 105–130.

[48] G. Budzik, The use of the rapid prototyping method for the manufacture and examination of gear wheels, in: M. Hoque (Ed.), Advanced Applications of Rapid Prototyping Technology in Modern Engineering, InTech, Croatia, 2011 (ISBN: 978-953-307-698-0).

[49] I. Gibson, D. Rosen, B. Stucker, Additive Manufacturing Technologies: 3D Printing, Rapid Prototyping, and Direct Digital Manufacturing, second Ed., Springer Science and Business Media Pvt Ltd, New York, 2015.

[50] P.J. Bártolo (Ed.), Stereolithography: Materials, Processes and Applications, first ed., Springer, New York, USA, 2011.

[51] H. Kodama, A scheme for three-dimensional display by automatic fabrication of three-dimensional model, IEICE Trans. Electron. (Japanese Edition) J64-C (4) (1981) 237–241.

[52] H. Kodama, Automatic method for fabricating a three-dimensional plastic model with photo-hardening polymer, Rev. Sci. Instrum. 52 (11) (1981) 1770–1773.

[53] X. Zhang, X.N. Jiang, C. Sun, Micro-stereolithography of polymeric and ceramic microstructures, Sens. Actuators 77 (1999) 149–156.

[54] M. Yoshinari, C. Weiwu, K. Soshu, Smart processing of 3-D micro ceramic devices by CAD/CAM micro-stereolithography and sintering, Trans. JWARI 36 (2) (2007) 57–60.

[55] L. Lu, J. Fuh, Y.S. Wong, Laser-Induced Materials and Processes for Rapid Prototyping, first ed., Springer Science & Business Media, New York (USA), 2001.

[56] P. Regenfuss, A. Streek, L. Hartwig, S. Klotzer, T.H. Brabant, M. Horn, et al. The performance of laser micro sintering with different material classes, in: Proceedings of the 31st Rapid Prototyping Symposium, 2007.

[57] H. Exner, M. Horn, A. Streek, F. Ullmann, L. Hartwig, P. Regenfuss, et al., Laser micro sintering: a new method to generate metal and ceramic parts of high resolution with sub-micrometer powder, Virtual Phys. Prototyp. 3 (1) (2008) 3–11.

[58] R. Udroiu, N.V. Ivan, Rapid prototyping and rapid manufacturing applications at Transilvania University of Brasov, Bull. Transilvania Univ. Brasov Ser. I: Eng. Sci. 3 (52) (2010) 2010.

[59] S. Somiya (Ed.), Handbook of Advanced Ceramics: Materials, Applications, Processing and Properties, second ed., Academic Press Inc, Waltham (USA), 2013.

[60] https://www.additively.com/en/learn-about/3d-printing-technologies.

[61] http://www.stratasys.com/3d-printers/technologies/polyjet-technology.

[62] U. Berger, Aspects of accuracy and precision in the additive manufacturing of plastic gears, Virtual Phys. Prototyping 10 (2) (2015) 49–57.

[63] R.E. Williams, S.N. Komaragiri, V.L. Melton, R.R. Bishu, Investigation of the effect of various build methods on the performance of rapid prototyping, J. Mater. Process. Technol. 61 (1-2) (1996) 173–178.

[64] D.T. Pham, S.S. Dimov, Rapid Manufacturing, Springer-Verlag London Limited, London, 2001.

[65] P.M. Pandey, K. Thrimurthullu, N.V. Reddy, Optimal part deposition orientation in FDM using multi-criteria GA, Int. J. Prod. Res. 42 (19) (2004) 4069–4089.

[66] C.C. Cheng, S.Y. Yang, D. Lee, Novel real-time temperature diagnosis of conventional hot-embossing process using an ultrasonic transducer, Sensors 14 (2014) 19493–19506.

[67] Y. Liu, W. Liu, Y. Zhang, D. Wu, X. Wang, A novel extrusion microns embossing method of polymer film, Mod. Mech. Eng. 2 (2012) 35–40.

[68] M. Worgull, in: J. Ramsden, W. Andrew (Eds.), Hot Embossing: Theory and Technology of Micro-Replication, first ed., Elsevier, UK, 2009, pp. 137–177.

[69] C.C. Kuo, B.C. Zhuang, A simple and low-cost method to the development of a micro-featured mold insert for micro-hot embossing, Indian J. Eng. Mater. Sci. 21 (2014) 487–494.

[70] F. Omar, Hot embossing process parameters: simulation and experimental studies, Ph.D. Thesis, Institute of Mechanical and Manufacturing Engineering, Cardiff School of Engineering, Cardiff University, Cardiff, Wales (UK), 2013.

[71] www.mems-exchange.org/capabilities/embossing.

[72] A. Feuerhack, D. Trauth, P. Mattfeld, F. Klocke, Fine blanking of helical gears, in: A.E. Tekkaya, W. Homberg (Eds.), 60 Excellent Inventions in Metal Forming, Springer, Heidelberg, 2015, pp. 187–192.

[73] Y. Shan, S. Yan-Li, Z. Mei, Effects of parameters on rotational fine blanking of helical gear, J. Cent. South Univ. 21 (2014) 50–57.

[74] S. Yang, L. Hua, Y. Song, Numerical investigation of fine blanking of a helical gear, Appl. Mech. Mater. 190–191 (2012) 121–125.

[75] V. Saile, in: V. Saile (Ed.), Introduction: LIGA and Its Applications, Advanced Micro and Nanosystems, vol. 7, Wiley-VCH Verlag GmBH and Co. KGaA, Weinheim, 2009, pp. 1–10.

[76] F. Kreith, R. Mahajan, MEMS Design and Fabrication, second ed., CRC Press, New York, USA, 2006.

[77] http://www.scme-nm.org/index.php?option=com_docman&task=cat_view&gid=84&Itemid=226.

[78] www.mems-exchange.org/MEMS/fabrication.html.

[79] T.R. Hsu, MEMS and Microsytems: Design and Manufacture, twentieth ed., McGraw Hill Education Private Limited, New Delhi (India), 2002.

[80] N.P. Mahalik, MEMS, (thirteenth ed.), McGraw Hill Education Private Limited, New Delhi (India), 2014.

[81] S. Debnath, M.M. Reddy, Q.S. Yi, Environmental friendly cutting fluids and cooling techniques in machining: a review, J. Cleaner Prod. 83 (2014) 33–47.

[82] E. Kuram, B. Ozcelik, E. Demirbas, Environmentally friendly machining: vegetable based cutting fluids, in: D.J. Paulo (Ed.), Green Manufacturing Processes and Systems, Springer, Berlin Heidelberg, 2013, pp. 23–47.

[83] K. Joyce, Producing green gears downloadable at http://www.machinery.co.uk, 2004.

[84] T. Tokawa, Y. Nishimura, Y. Nakamur, High productivity dry hobbing system, Tech. Rev. Mitsubishi Heavy Ind. Ltd 38 (2001) 27–31.

[85] http://www.mhi.co.jp/technology/review/pdf/e523/e523009.pdf.

[86] E.P. Kovar, Dry gear hobbing, Gear Technol. Magazine, July/August (1995) 39–41.

[87] D. Fratila, Evaluation of near-dry machining effects on gear milling process efficiency, J. Cleaner Prod. 17 (2009) 839–845.

[88] D. Fratila, A. Radu, Modeling and comparing of steady thermal state at gear milling by conventional and environment-friendly cooling method, Int. J. Adv. Manuf. Technol. 47 (2010) 1003–1012.

[89] D. Fratila, Research of environment-friendly techniques influence on gear accuracy in context of sustainable manufacturing, in: Proceedings of the Romanian Academy: Series A, 14 (2013) 56–63.

[90] A. Filipovic, D.A. Stephenson, Minimum quantity lubrication (MQL) applications in automotive powertrain machining, Mach. Sci. Technol. 10 (2007) 3–22.

[91] W. Stachurski, Application of minimal quantity lubrication in gear hobbing, Mech. Mech. Eng. 16 (2012) 133–140.

[92] G. Zhang, H. Wei, Selection of optimal process parameters for gear hobbing under cold air minimum quantity lubrication cutting environment, in: Proceedings of the 36th International MATADOR Conference, Manchester, UK, 2010, pp. 231–234.

[93] X. Zhang, C. Xia, P. Chen, G. Yin, Comparative experimental research on cryogenic gear hobbing with MQL, Adv. Mater. Res. 479–481 (2012) 2259–2264.

[94] B.L. Tai, D.A. Stephenson, R.J. Furness, A.J. Shih, Minimum quantity lubrication in automotive powertrain machining, Proc. CIRP 14 (2014) 523–528.

Chapter 5

Conventional and Advanced Finishing of Gears

It is estimated that the global demand for gears exceed ten billion units annually in various industrial, commercial, military, home appliances, scientific instruments, and other applications. Such a significant demand is further exacerbated by the ever-increasing requirements for improvement in performance and cost reduction. This includes (1) enhancement in power transmission capability, surface finish, surface integrity, fatigue strength, operating performance, reliability, service life, and recycling; (2) decrease in vibrations, noise, harmful emission, price, disposal challenges, and unpredictable sudden failures; and (3) increased use and integration of intelligent and automated systems in the gear drives for failure prediction, operation control, data acquisition, and safety [1].

Unpredictable and/or catastrophic gear failures are usually classified depending on if they were lubrication related or not. Failures of gears due to overload and fatigue caused by bending are typically non-lubrication-related failures, whereas scuffing, wear (adhesive, abrasive, corrosive, and fretting type), and pitting and micropitting caused by Hertzian fatigue are significant lubrication-related gear failures. Prevention of gear failures demands careful consideration on design, operating environment and surface quality. Gear design determines the geometry, material, motion, lubricant characteristics, and static and dynamic forces on the gear tooth. Assessment of the surface quality of a gear involves evaluation of surface roughness parameters, deviations or errors in microgeometry and wear characteristics of the flank surfaces of its teeth. Poor surface finish may adversely affect the service life and operating performance of a gear. It also increases running noise particularly at higher operating speeds and may further also introduce transmission errors which may be exacerbated by errors in microgeometry and low wear resistance. A poor surface finish may also lead to micropitting and large-scale pitting which may result in sudden premature failure of the gear. Wear characteristics, microstructure, and fatigue strength may also significantly affect operating performance and service life. Therefore, optimum operating performance including noise levels and an extended service life require (1) accurate overall dimensions (low tolerances); (2) minimum surface roughness; (3) minimum errors in its microgeometry; and (4) higher hardness and

Advanced Gear Manufacturing and Finishing. DOI: http://dx.doi.org/10.1016/B978-0-12-804460-5.00005-5

wear resistance. The surface quality of the gear tooth may be enhanced appreciably by the use of an *appropriate* finishing process. It implies a finishing process that (1) improves surface quality and reduces errors in its microgeometry to minimize the running noise; (2) improves surface integrity of the gear to maximize its load carrying capacity; and (3) is able to perform flank modifications, such as flank crowning, profile crowing, providing relief at tip and root, modifications in topology of gear flank surfaces, and other desired minor corrections in gear tooth form.

The cost attributed to finishing typically contributes 15%−25% toward its total manufacturing cost [1,2]. Finishing costs increases rapidly if good surface finish, especially below 1 μm roughness, is required. Consequently, the significant demand and use of gears has led to the development of different finishing processes that may be grouped either as (1) conventional processes such as gear shaving, gear grinding, gear honing, gear lapping, gear burnishing, and gear skiving; or (2) advanced processes such as electrochemical honing (ECH), abrasive flow finishing (AFF), water-jet deburring (WJD), electrolytic deburring (ED), thermal deburring, brush deburring, vibratory superfinishing, black oxide finishing, and selected hybrid processes such as ultrasonic-assisted abrasive flow finishing (US-AFF), etc. The selection of a specific gear finishing process is a function of various aspects such as the initial process used to manufacture the gear in question, hardness of gear material, need and type of case hardening process, roughness value of the unfinished gear, type and size of gear, and production volume. For example, small gears manufactured by the cold-rolling process generally do not require any finishing [2]. Cold-rolled gears requiring case hardening may require limited finishing. Gears manufactured by near-net-shape manufacturing processes such as die-casting, powder metallurgy, and extrusion, etc. usually do not require significant finishing. Machined and subsequently hardened gears may need finishing for improving the microgeometry and surface finish.

5.1 CONVENTIONAL FINISHING PROCESSES FOR GEARS

The most common conventional processes for gear finishing are gear shaving, gear grinding, gear honing, gear lapping, gear burnishing, and gear skiving. Cylindrical (i.e., spur and helical) gears can be finished by any of these processes, but only grinding and lapping are appropriate for finishing conical gears. Table 5.1 presents a comparative summary of these processes highlighting their capabilities, advantages, limitations, and applications. This is followed by an introduction of each method in some details according to their working principle, types, cutting tools, and other relevant details.

TABLE 5.1 Capabilities, Advantages, Limitations, and Applications of Conventional Processes for Finishing the Gears

Process	Capabilities and Advantages	Limitations	Applications
Gear shaving	Can correct selected errors in tooth profile, lead angle, and runout thereby reducing running noise Able to yield an average surface roughness of 0.4 μm Routinely produce to DIN class 6 or 7 quality Can modify profile which increases load-carrying capacity of the gears May reduce thermal distortions associated with heat treatment Economical and fast	Used for finishing gears up to a maximum hardness of approximately *40 HRC* only A step mark is left on the involute profile which may cause excessive wear, noise, and vibrations Comparatively removes more material from the pitch surface which may deteriorate surface finish and may affect transmission quality	Used for finishing external cylindrical gears and *worm wheels* of moderate size Finishing of small gears produced in high volumes Gears with minor profile and lead errors Gears with minor thermal distortions To remove gear hobbing irregularities including scallops For crowning and tapering of gears
Gear burnishing	Improves surface integrity and fatigue life of the gears Higher production rates than other conventional finishing processes due to typical finishing times of 5–10 seconds Can finish both external and internal gears Limits burrs and nicks from the gear teeth Longer life of burnishing dies and reduced maintenance cost Well suited for mass production	Can be used for unhardened gears only Being a localized cold-working process, it may result in localized surface residual stresses and nonuniform surface characteristics Cannot improve tooth position, tooth profile, lead, spacing, and concentricity Burnishing dies are costlier than shaving and grinding cutters May induce substantial damage if a large section of the gear tooth is burnished	For finishing both internal and external helical gears For improving profile surface finish and therefore reducing running noise For improving fatigue strength of gear teeth

(*Continued*)

TABLE 5.1 (Continued)

Process	Capabilities and Advantages	Limitations	Applications
Gear skiving	Rough machining and finishing in one clamping operation Faster than gear shaping Increased productivity as no extra loading time needed Fast and continuous machining of internal and external gears Both straight and helical gears can be cut/finished	Not exclusively a finishing process therefore limited surface quality achievable only Being essentially a rehobbing process, it suffers from similar limitations as associated with gear hobbing Can induce mechanical damage in the form of microcracks, hardness alternation, and localized plastic deformation Not very effective in reducing microgeometry errors	To correct the cumulative spacing and concentricity errors For finishing internal and external gears More suitable for medium volume production Finishing coarse and fine pitch gears Attractive alternative finishing process for many hardened gears
Gear grinding	Can finish high strength and hardened gears Only productive process to finish conical gears Fast, accurate and capable of providing good finish Can correct most microgeometry errors with an appropriately shaped grinding wheel Most frequently used gear finishing method	Expensive and complex May results in two significant undesirable effects, i.e., grinding burns and transverse grind lines on the ground surfaces which may increase noise and vibration Heat input due to grinding may deteriorate surface integrity including formation of fine cracks, thermal distortion, and uneven stress distribution Not effective for internal, multi-joint and large-size gears Requires frequent redressing of wheel	For finishing high strength and/or hardened spur, helical and conical gears For finishing gears with large pressure angles, face widths or modules For finishing gears requiring high accuracy in profile and lead For mixed volumes and frequent changeovers

Gear lapping	Can correct small errors in hardened gears including helix angle, spacing, involute profile, and eccentricity May significantly improve gear tooth contact May significantly improve wear properties of gear teeth Induces less damage than grinding and may even alleviate some subsurface damage Excellent parallelism and flatness Provides good dimensional accuracy and tolerance $<2.5\ \mu m$	Only able to improve small errors produced during manufacturing and heat treatment. These include involute profile, helix angle, tooth spacing, and concentricity Finishes pair of mating gears that are not interchangeable with other gear pair Slow and costly process that may adversely affect involute profile if used for long durations	For finishing high strength and/or hardened spur, helical, bevel, spiral bevel, and hypoid gears For finishing shaved and *hardened* gears For finishing a pair of mating gears requiring high dimensional accuracy, tolerance $<2.5\ \mu m$ with excellent surface quality
Gear honing	Produces cross-hatch lay pattern which helps in lubrication retention thereby lowering friction and wear Improves microgeometry, dimensional accuracy, roughness, lay pattern, and surface integrity Low-temperature process with reduced surface hardening and microstructural changes No significant surface residual stresses, heat cracking, or burn spots Economical	Honing gear tool has limited life Honing time increases with increasing honing tool wear	For finishing high strength and/or hardened internal and external gears Used for gears with high contact ratios, low pressure angles and longer addendum For crowning of gears Typically used for minor corrections of gear microgeometry For removing nicks and burrs Improving resistance to wear and pitting

5.1.1 Gear Shaving

Gear shaving is a fine finishing process that applies a shaving cutter to remove gear workpiece material in the form of hair-like chips with lengths typically in the range of 100–400 μm and thickness between 50 and 150 μm. Because of the low cutting speed thermal loading on the cutting edges are low [3]. The performance of this process depends on the type, geometry, and material of the shaving cutter, the shaving allowance, fundamental process parameters associated with the shaving, shaving machine tool associated parameters, workpiece gear material, and its geometry. This process is typically used in finishing of automotive gears and gears used in construction machinery due to its cost efficiency.

Shaving cutters are classified according to their basic geometry, i.e., (1) helical shaving cutter; (2) rack type shaving cutter, and (3) worm shaving cutter (see Fig. 5.1). These cutters are equipped with serrated cutting edges with a negative clearance angle. The selection of shaving cutter material is a function of the gear workpiece hardness. Different grades of hardened high-speed steel (HSS) are generally used in many gear shaving applications, whereas carbides are used for shaving of the gears made of difficult-to-finish materials. To obtain a good finish and acceptable accuracy, it is important to allow sufficient extramaterial thickness on the gear workpiece. Jain and Petare [2] have suggested standard shaving allowances for gears of different sizes.

Gear shaving may be classified according to cutter type and relative motion (to gear workpiece), i.e., (1) rotary shaving and (2) rack shaving.

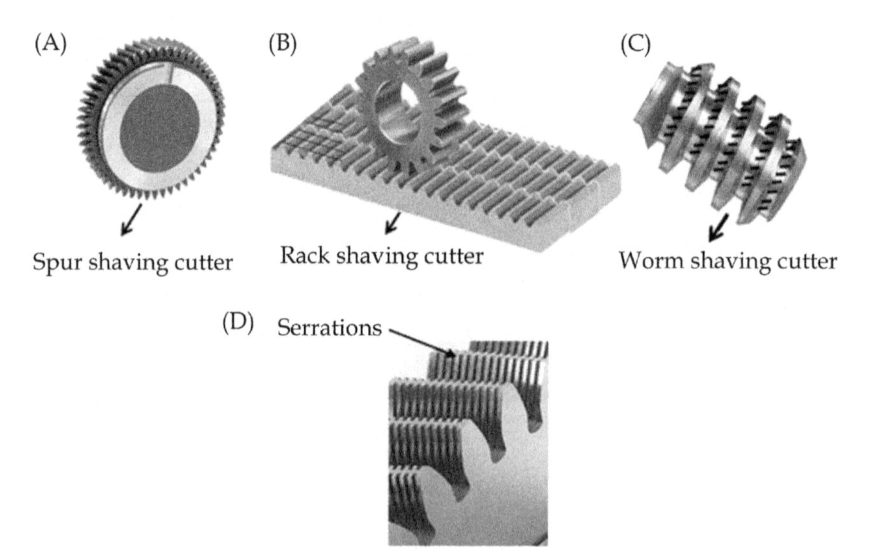

FIGURE 5.1 Examples of (A) spur shaving cutter; (B) rack shaving cutter; (C) worm shaving cutter; and (D) enlarged view of the serrations on a shaving cutter *[3]*.

- *Rotary shaving* utilizes a shaving cutter in the form of a helical gear (Fig. 5.1A) with serrated teeth ground in such a way that its teeth profile has a conjugate shape with respect to the workpiece gear to be shaved [4]. Rotary gear shaving may be performed in four different ways depending on the relative motion between workpiece gear and shaving cutter [2]. These are (1) conventional or axial gear shaving (Fig. 5.2A); (2) tangential or under pass gear shaving (Fig. 5.2B); (3) plunge gear shaving; and (4) diagonal gear shaving.
- *Rack shaving* utilizes a reciprocating rack type shaving cutter (depicted in Fig. 5.1B). Feed is incrementally applied at the end of each stroke. Gear sizes up to approximately 150 mm (6 in) are finished by this process.

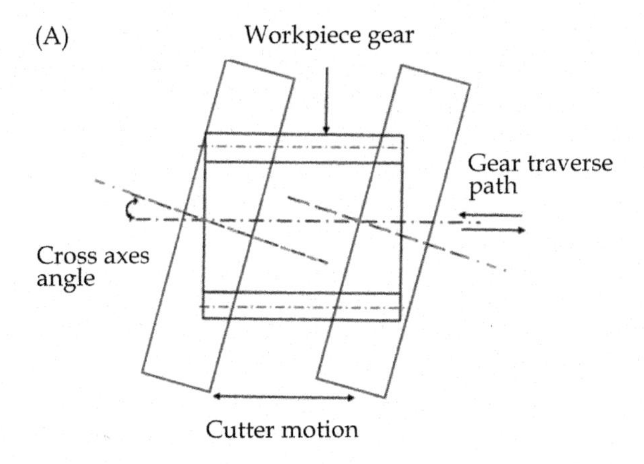

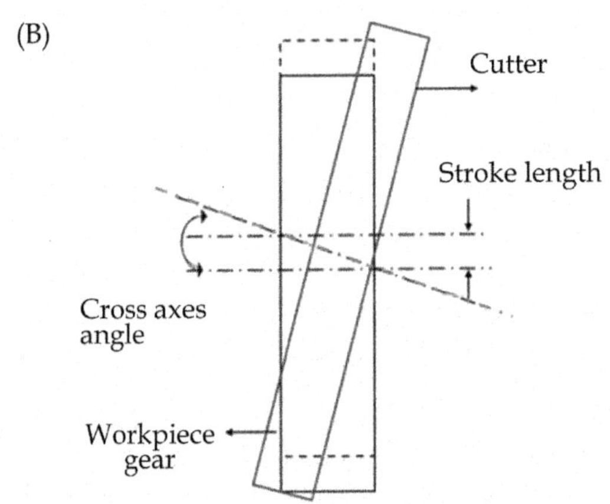

FIGURE 5.2 Concept of (A) conventional or axial gear shaving and (B) tangential or underpass gear shaving [5].

In *conventional* or *axial gear shaving*, the shaving cutter rotates while meshed with the workpiece gear in either a parallel or crossed axes arrangement (Fig. 5.2A). The cross-axes angle refers to the angle between the shaving cutter and workpiece gear rotational axis. It is typically similar to the associated helix angles. Different cross-axes angles are possible by appropriate meshing between the shaving cutter and workpiece gear but typically angles of between 10° and 15° results in the best surface finish and dimensional accuracy. Higher cross-axis angles results in increased material removal, but it reduces the contact zone width and consequently affects the cutting action. The shaving cutter is reciprocated along the axis of the workpiece gear to enable cutting edge penetration. The stroke length is generally adjusted such that it is at least a few millimeters more than the face width of the workpiece gear to ensure complete finishing. As finishing progresses, the center distance between the shaving cutter and workpiece gear is gradually reduced to maintain appropriately close meshing and an appropriate contact pressure. The finer the cut, the lower is the contact pressure required between cutter and workpiece thus reducing cold working of the workpiece gear material. Direct contact between the shaving cutter and the gear root fillet should also be avoided due to its adverse effect on the involute profile accuracy. The rotational direction of the shaving cutter is changed from clockwise to anticlockwise and vice-versa. This process is especially effective for the gears with extended face widths. It typically reduces the surface roughness of the flank surfaces of the gear teeth and single and adjacent pitch errors. It cannot however remove or decrease cumulative pitch or index errors and may in fact actually increase these due to the formation of grooves on the gear teeth flank surfaces [6].

In the *tangential or underpass gear shaving* process, the workpiece gear is reciprocated in the radial direction while rolling with the shaving cutter as shown in Fig. 5.2B. It requires that the shaving cutter be wider than the face width of the workpiece gear. The pitch surface of the shaving cutter is of hyperboloid form with a concave curvature to ensure appropriate contact across the full face width of the workpiece gear. The length of the reciprocating stroke is smaller and the wear of the cutter is more uniform when compared to conventional and diagonal gear shaving. This results in higher productivity and increased cutter life. It is particularly useful for finishing gears with a shoulder.

In *plunge gear shaving*, the workpiece gear is plunged into a shaving cutter with a hollow tooth form with a pitch surface of hyperboloid shape [7]. This eliminates the necessity for reciprocating motion of the shaving cutter and/or workpiece gear. It also results in line contact between the shaving cutter and workpiece gear tooth surfaces which is unlike other versions of gear shaving that mostly involves point contact. This does however necessitate the use of wider shaving cutters with special tooth flank geometry and crossed-axes angles limited to 10°–20°. This implies a short stroke length and consequently reduced shaving time in the order of 10 seconds for a gear of a typical automobile gearbox. Other than high productivity, this process also provides good accuracy, enhanced life of the shaving cutter, and simple design of the machine tool. Hence, this process is eminently suitable for shaving gears for large production volumes.

Diagonal gear shaving implies a special relationship between the face widths of the shaving cutter and workpiece gear with the diagonal traverse angle varying between 30° and 60°. A wide workpiece gear face width along with a narrow shaving cutter results in a small diagonal traverse angle. This leads to uniform wear of the shaving cutter that usually implies an increased life. It is therefore used mainly for finishing gears in medium to high-production volumes.

5.1.2 Gear Grinding

Gear grinding is used to finish cylindrical and conical gears of high strength and/or hardened materials with Rockwell C scale harnesses in excess of approximately 40 by removing small amounts of material from the flank surfaces of the workpiece gear by abrasive action. The grinding wheel (cutting tool) consists of special abrasive grains that are imbedded in an appropriate matrix. Alumina oxide (Al_2O_3), silicon carbide (SiC), and cubic boron nitride (CBN) are the most commonly used abrasives grains. Gear grinding has the ability to correct thermal distortions induced typically during case hardening and may also enhance the surface finish and microgeometry of the gears thereby improving the overall quality as applicable to various engineering applications. The abrasive grains of the grinding wheel are in contact with the workpiece flank surface only for a fraction of the rotation during which time they generate chips in a cutting process. Three distinct phenomena namely rubbing, ploughing, and cutting occur during this period. Actual metal removal (cutting) only occurs when the force exerted by the grinding wheel on the workpiece gear exceeds a threshold value (known as *threshold force*). Beyond the threshold value rubbing and ploughing transition to cutting. Typically gear finishing by grinding may be subdivided into two main categories namely (1) nongenerative or form grinding of gears and (2) generative grinding of gears. Fig. 5.3 depicts the fundamental differences between these categories.

- **Nongenerative or form grinding of gears**: It involves gear finishing by using a specially designed grinding wheel with a circumference profiled to correspond with the involute form of a particular type of gear geometry. This implies that a unique grinding wheel is required for different modules, pressure angles, and number of teeth. A specially designed grinding wheel with unique profile is located in the area between adjacent teeth of the workpiece gear as shown in Fig. 5.3B. Grinding may be

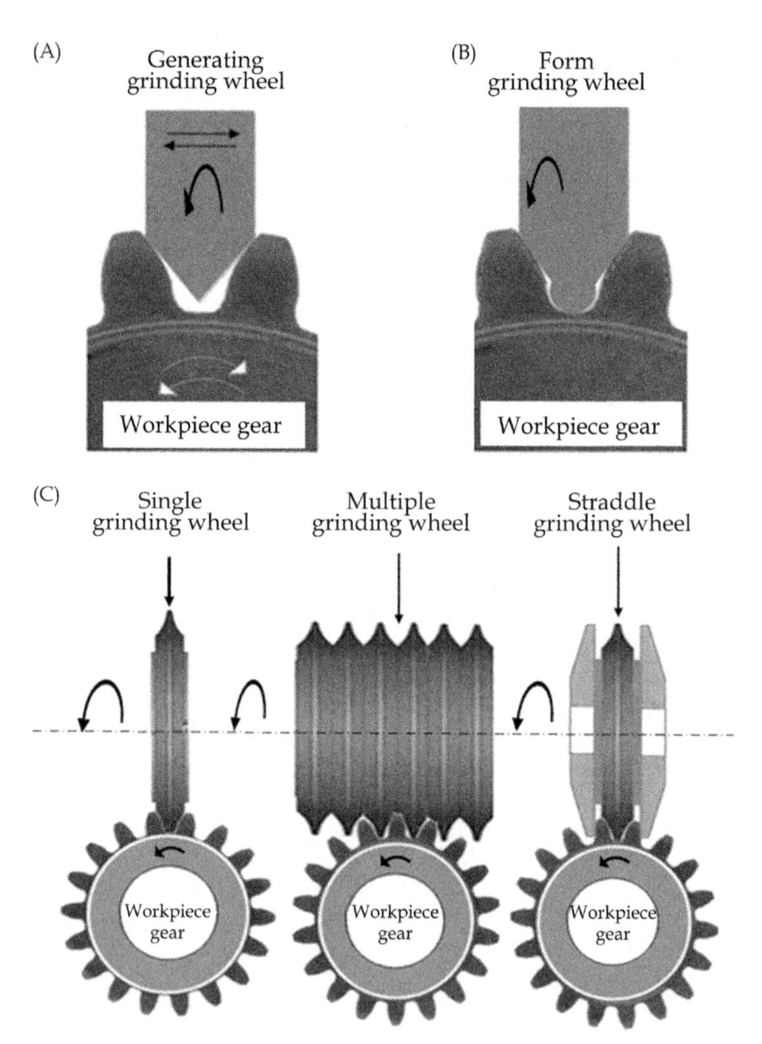

FIGURE 5.3 Finishing of a gear by (A) generative grinding; (B) nongenerative or form grinding; and (C) different forms of the grinding wheels used in nongenerative grinding *[3]*.

conducted with a single form, multiple form or a straddle wheel (see Fig. 5.3C). Rotation of the form wheel ensures generation of the corresponding tooth profile on the workpiece gear. Indexing is applied to ensure that all teeth are finished. This does however make the process slow and less accurate. This process is simple and any complex form of a gear tooth can be finished by designing and manufacturing an appropriately shaped grinding wheel. Nongenerative grinding can be used for finishing spur gears, single helical gears, straight bevel gears, and worm and worm wheels. Two further categories may be identified according to the type of abrasive used, i.e., (1) *ceramic form grinding* which uses a form grinding wheel containing ceramic abrasives such as Al_2O_3, and SiC to grind gears with hardness in excess of 60 HRC; and (2) boron *form grinding* which contains cubic boron nitride (CBN) as abrasives.

- **Generative grinding of gears**: This implies finishing of various shaped gears by using a generic rotating grinding wheel that generates the required profile by synchronous rolling of the workpiece gear (see Fig. 5.3A). Different wheel shapes may be utilized including a threaded wheel, dish-shaped wheel, cup-shaped wheel, and rack-tooth worm wheel. Indexing is required to finish the full complement of teeth. It is commonly used for gears with modules between 0.5 and 10 mm.

The working principle of generative grinding by *threaded wheel* is similar to gear hobbing except that it uses a threaded grinding wheel. The workpiece gear and grinding wheel rotate about their axes with their relative speed ratio being a function of the workpiece gear teeth number and the number of thread starts on the wheel. The indexing mechanism largely determines the eventual quality of the gears finished by this process. It can be used for finishing external spur and helical gears. Generative grinding by *cup-shaped wheel* uses a grinding wheel with a cup-shaped profile on its circumference whose outer edges meshes with the workpiece gear and are chamfered at an angle equal to the pressure angle of the workpiece gear. The rotating grinding wheel finishes those flank surfaces of the two adjacent teeth to which its outer edges are in contact. The workpiece gear requires indexing to finish all the teeth. It can be used to finish spur and helical gears with pitch circle diameters up to approximately 450 mm.

Grinding by *dish-shaped wheels* involves two dish-shaped grinding wheels arranged either parallel to each other or at an inclination angle (typically 15°–20°) to finish the gear by generation. It is typically used when high-accuracy gear teeth profiles and longitudinal modification are required at high productivity levels. Its biggest limitation is a nonuniform geometric transition between the gear tooth root and flank.

The grinding wheel is in the shape of a conventional worm wheel and may be CBN plated. Helical gears are finished by aligning the slide of the machine tool with the helix angle of the workpiece gear. The quality of the finished gear depends primarily on the profile of the grinding wheel, accuracy of grinding wheel dressing process, and consequently on the diamond wheel accuracy used for the dressing.

Rotational speed, grinding wheel wear, grinding fluid, and grinding time are important parameters which affect performance of the gear grinding process. An increase in the grinding wheel speed (all other parameters being constant) reduces the grinding force, reduces roughness and increases the life of the redressed grinding wheel all at a lower specific energy consumption. Grinding wheel wear or the so-called *G*-ratio usually refers to the volume ratio of workpiece material removed relative to the grinding wheel wear loss. Correctly performed grinding may be a self-sharpening process in which the wear of the cutting edges of the abrasive grains may lead to an increased cutting force to such an extent that either the abrasive particle is fractured and new sharp cutting edges are revealed or the particle is removed in totality from the matrix thereby exposing new abrasive grains. The use of appropriate grinding fluids facilitates heat and chip removal which enhances grinding wheel life, gear tooth finish and minimizes the occurrence of grinding burns on the gear flank surfaces. Typically, mineral-based sulfochlorinated and/or sulfurized oils are used. Plain soluble-oil emulsion solutions are also used with satisfactory results in various gear grinding applications [8]. The productivity of gear finishing by grinding is a function of the time required for rough grinding, finish grinding, and dressing of the grinding wheel. Grinding wheel wear is unavoidable and therefore frequent replacement is required. Other nonproductive activities such as wheel dressing (restoration of sharp cutting edges of the abrasive grains), wheel truing (wheel balancing), and wheel profiling (restoring circumferential profile of the wheel) is also required and has a negative effect on productivity.

5.1.3 Gear Honing

Gear honing is used to finish gears made of high strength and/or hardened material using an abrasives-impregnated helical gear as the honing tool that closely meshes with the workpiece gear arranged in a cross-axes alignment. The honing gear drives the workpiece gear at high speed up to approximately 300 m/min and at the same time the workpiece gear is reciprocated along its axis thereby finishing the entire face width. Superimposition of the rotary and reciprocating motion yields honed flank surfaces of with a cross-hatch lay pattern where the orientation depending on the ratio of the two motions. This lay pattern facilitates improved lubricating oil retention in axial and circumferential directions, reduces frictional resistance, and a more uniformly distribution of the load. Typically, 0.013–0.05 mm is removed from the workpiece. An appropriate cutting fluid is used to flush out the chips from the finishing zone and to facilitate heat removal. Gear honing removes nicks, burns, and minor irregularities from the active profile of the workpiece gear teeth, yields higher dimensional accuracy and better finish of gear flank surfaces, and improves noise and wear characteristics of the workpiece gears [9].

Gear honing may be done either externally or internally depending on the arrangement of the honing gear and workpiece gear. In the *external honing process*, the workpiece and honing gears are meshed in a cross-axes alignment with the center distance being maintained as shown in Fig. 5.4A. The gear tooth is finished from the root to the tip with the direction of rotation of the honing gear reversed for finishing the second flank. The *internal honing process* uses a large-size internal helical gear as honing tool that meshes

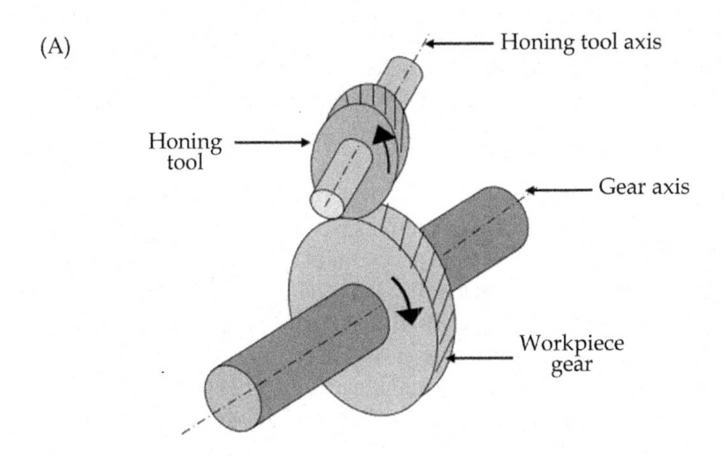

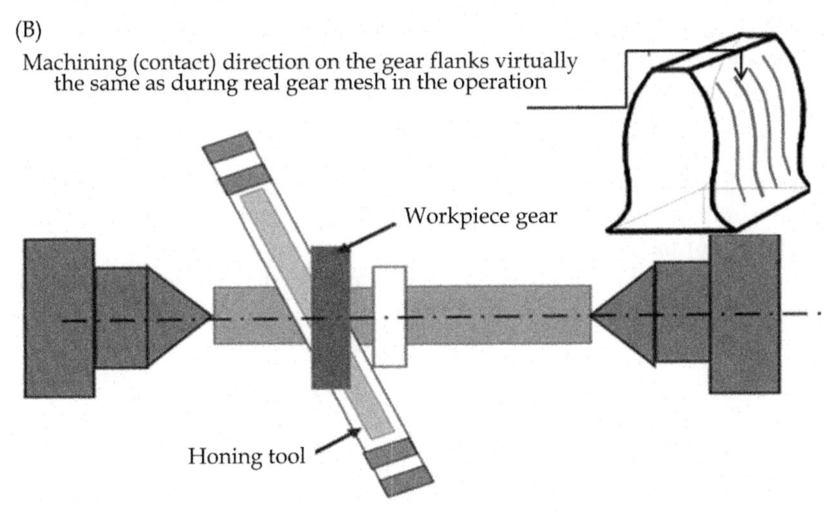

FIGURE 5.4 Types of gear honing (A) external honing of a single helical gear (Source: from [3]. Elsevier © 2017, reprinted with permission) and (B) internal honing of an external spur gear *[9]*.

with the workpiece gear in a cross-axes arrangement fixed between the centers of the honing machine as depicted in Fig. 5.4B. The alignment of the honing gear axis can be changed to achieve the required crossed-axes angle. The gear honing machine tool reciprocates the workpiece gear. Internal honing offers certain advantages over external honing including: (1) it provides a higher traverse contact ratio than that of external honing which results in a more balanced contact or honing pressure, larger honing stroke, better quality of the workpiece gear with smaller adjacent pitch errors, accumulated pitch errors, and runout; (2) it provides higher accuracy; and (3) it provides higher stability than external honing thus offering increased fracture resistance. According to the tolerance specification, the workpiece gear can be finished either by a *zero-backlash* gear honing process if its dimensional tolerances are within the commercial ranges or by *constant-pressure* gear honing if its tolerances are outside commercial ranges.

The surface finish of a honed gear depends on (1) abrasive size and type; (2) applied honing pressure; (3) rotational speed; and (4) speed and length of the reciprocation. The honing gear material should be of sufficient strength so that its deformation is minimized under load while finishing the flank surfaces of the workpiece gear. Generally, it consists of abrasive grains contained/bonded within a matrix of resinoid, vitrified, or metallic type materials. The associated bond between the abrasive grains and matrix should be sufficiently strong to retain the abrasive grains but still allow a modicum of deformation. Excessively strong bonding may deteriorate finish and quality of the workpiece gear instead of improving it. Alumina, SiC, CBN and diamond are the most commonly used abrasives for honing. The size of the abrasive grains typically range from 60 to 500 mesh number but actual size depends on the circular pitch of workpiece gear, desired surface finish and material removal rate (MRR). The honing gear type is a function of the workpiece gear material, type of honing, application requirements, and capabilities of the gear honing machine tool. Generally, the diameter of honing gears range from approximately 100 (for internal gears) to 350 mm and face width ranges between 10 and 50 mm.

5.1.4 Gear Lapping

Gear lapping is a slow-speed (<80 rpm) low-pressure process used to finish cylindrical and conical gears made of high strength and/or hardened material by an abrading action of an abrasive-laden lapping compound continuously supplied under pressure. Lapping of cylindrical gears (i.e., spur, single helical, double helical and herringbone gears) involves meshing the workpiece with a gear-shaped lapping tool. The sliding velocity decreases from a maximum value at the root of a tooth to zero at the pitch line and then increases again to a maximum value at the tip of the tooth. An auxiliary sliding action in the axial direction of the cylindrical gears is required to compensate for the lack of sliding in the vicinity of the pitch line. This ensures a uniform lapping action across the entire tooth profile. The lapping tool is typically made from a material that is softer than the workpiece gear material to minimize its wear due to the abrasive grains which will then preferentially rather embed themselves into its surface instead of moving across it while removing workpiece material. The most commonly used lapping tool material is fine-grained cast iron.

Lapping of conical gears (straight bevel, spiral bevel, and hypoid gears) involve running the mating gears together, as shown in Fig. 5.5, subject to a controlled load. In both the cases, the larger gear rotates freely while being driven by the smaller gear. There is no need to provide a finishing allowance on the mating conical gears because both mating gears finish simultaneously. Typically, the lapping process is preceded by both gears initially ground by hand to remove burrs and nicks. The lapping medium appears as a chalky paste comprising of abrasive grains with sizes between 300 and 900 mesh size (i.e.,particle size approximately ranging from 17 to 51 μm) and a carrier fluid. The most commonly used abrasives are aluminum oxide, SiC, boron carbide and diamond powder. Typically, spiral bevel gears are lapped with abrasive grains of 280 mesh size, whereas hypoid gears are typically lapped with a 400 mesh size. Coarser abrasive grains (i.e., larger size) are used for gears with a coarse pitch and finer abrasive grains are used for gears of finer pitch. The most common carrier fluid that are used include oil, grease, water, and kerosene. Lapping works by removing minute amounts of material because of the relative motion of the rotating mating gears. The rotation direction is changed thereby imparting a good surface finish, high accuracy, and the appropriate contact pattern. The success of conical gear lapping is a function of essentially four lapping parameters. These include lapping pressure, abrasive grain size, abrasives concentration in the lapping medium, and the speed. Noise level reduction is typically an important outcome of effective lapping [10].

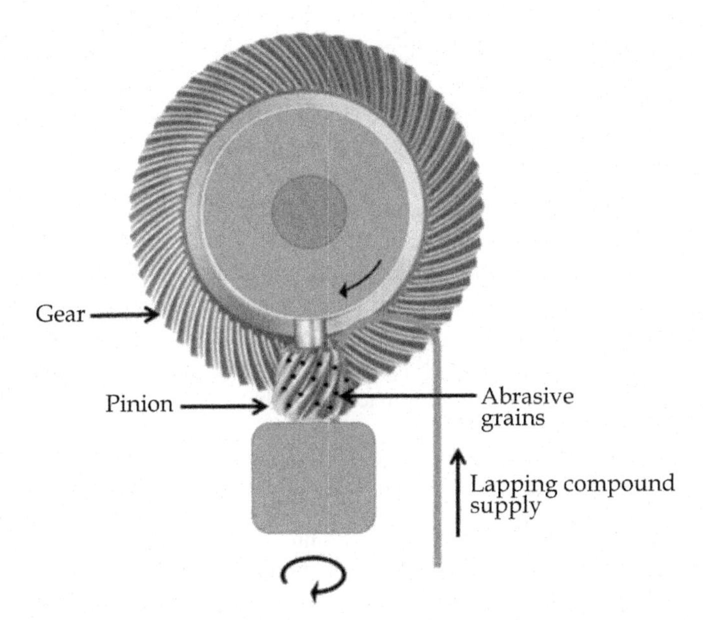

FIGURE 5.5 Finishing of a spiral bevel gear by the gear lapping process *[3]*.

5.1.5 Gear Burnishing

Gear burnishing is a process that modifies the gear surface and geometry by cold forming instead of material removal as associated with the other finishing processes. Low strength and/or unhardened gears are meshed accurately under pressure with ground, polished, and hardened master gears, also referred to as burnishing dies, of high accuracy and surface finish. This causes smearing off of the minute irregularities on flank surfaces of the workpiece gear teeth by cold work that improves surface finish but may also be beneficial to other surface integrity descriptors. Typically, a single burnishing gear is mounted on a heavy duty gearhead and meshed with the workpiece gear. Two idle gears are used to provide support (see Fig. 5.6A). The workpiece gear is mounted such that it is free to rotate. The burnishing gear is driven and drives the whole assembly by appropriately close meshing with the workpiece gear. The direction of rotation of the burnishing gear is alternated to facilitate burnishing of both the leading and trailing side of the workpiece gear teeth. A lubricant can be used to reduce abrasive wear of the workpiece gear and to improve the surface quality.

Double gear burnishing works by sandwiching the workpiece gear between an upper and a lower burnishing gear as depicted in Fig. 5.6B. Typically, one of the two burnishing gears is hydraulically controlled to provide a feed force. The workpiece gear is rotated at slightly different speed than that of the upper burnishing gear. The upward feed provided by the hydraulically controlled burnishing gear controls the finishing of the workpiece gear.

The direction of sliding changes during engagement of the burnishing gear with the workpiece gear. Typically, sliding occurs along a line of action from the top of the burnishing gear tooth toward its pitch point on the approach side then onward from the pitch point toward the tooth root on the recession side. This induces a compression stress in the workpiece gear surface from root toward pitch point on the approach side and from pitch point toward the tooth top on the receding side. More material is displaced on the receding side than the approach side in a ratio of approximately 3:1 [11]. The effectiveness and quality of the gear burnishing process is affected by the workpiece gear hardness, extent of plastic deformation imparted, and the finishing allowance. Burnishing gears are typically manufactured from HSSs with increased fatigue and impact strengths.

5.1.6 Gear Skiving

Gear skiving essentially implies rehobbing of a hardened gear using a carbide helical gear as cutter (unlike gear hobbing which uses a worm-like cutter) with a negative rake angle. Gear skiving is a highly efficient process due to its continuous generative machining action making it an effective

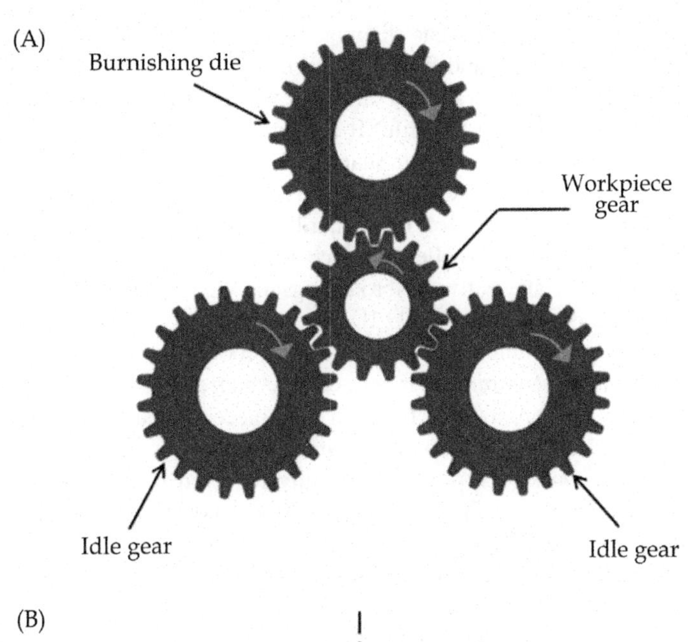

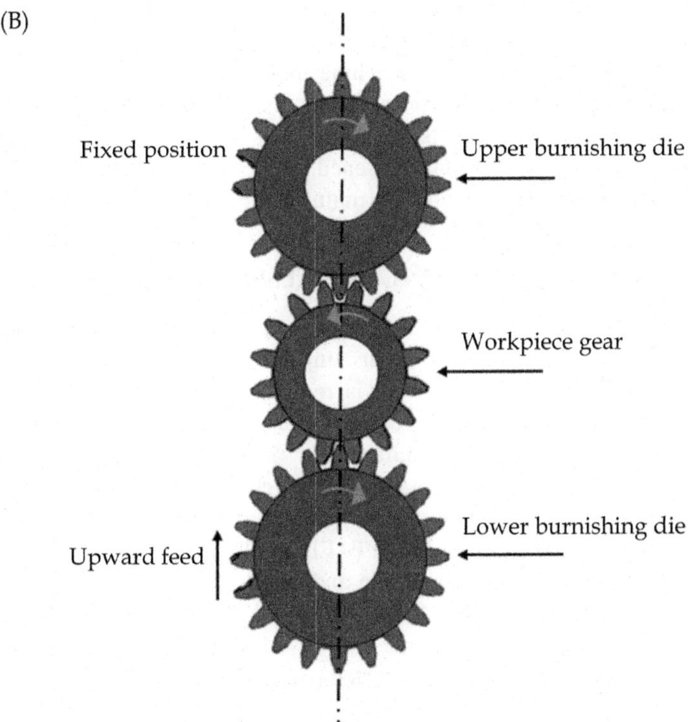

FIGURE 5.6 Schematic of (A) single gear burnishing and (B) double gear burnishing. *[3]*.

alternative to broaching and gear shaping. Its most significant advantage is that both rough machining and finishing can be done in one clamping operation of the workpiece gear. It may improve productivity 2–3 times while also significantly increasing tool life. It is useful for finishing internal gears manufactured by broaching or gear shaping. It is more suitable for medium volume production. Consequently, gear skiving has become an attractive alternative finishing process for many hardened gears because it provides both high quality and an excellent surface finish. For finishing of internal gears, it provides an alternative to broaching and gear shaping. Both coarse and fine-pitch gears can be skived [12].

A high-quality machine tool is necessary to produce quality skived gears. The most important requirement is an accurate and efficient automatic system that ensures synchronization between workpiece and cutter along with a highly accurate dividing (positioning) system that ensures that the finishing allowance on the workpiece gear is uniformly removed from both flanks of the workpiece gear teeth. Typically, an electronic noncontact system picks up a sequence of impulses from the rotating workpiece gear that is then subsequently processed to determine the angular position used to obtain proper alignment between workpiece and hob. Inaccurate alignment may lead to nonuniform material removal of the gear teeth. Gear skiving can be done on conventional and CNC hobbing machine tools provided they have adequate static and dynamic rigidity. A properly equipped, modern, high-quality hobbing machine tool may then be used not only to machine the gears in the soft state but also for skiving hardened gears of acceptable quality. This reduces capital costs associated with multiple machines [12,13].

5.2 ADVANCED FINISHING PROCESSES FOR GEARS

Table 5.1 illustrates the various limitations and nonoverlapping nature of the conventional finishing processes. This implies that in certain cases, the use of more than one finishing process is required, often with a suitable heat-treatment process incorporated between them, to achieve the desired gear quality and characteristics. Manufacturing productivity may be limited if vibration-free, low noise transmission, and extended service life (especially at higher transmission speeds) is to be guaranteed. In this context, advanced finishing processes such as ECH, AFF, ED, WJD, etc. have emerged as processes which may yield high precision and quality finishing of gears by virtue of certain unique capabilities. These advanced processes may offer certain unique advantages while at the same time have the potential to address some of the most significant limitations associated with conventional finishing.

5.2.1 Gear Finishing by Electrochemical Honing

Electrochemical honing (ECH) is a state-of-the-art superfinishing process that is basically the hybridization of mechanical honing and electrochemical finishing (ECF). If a pulsed direct current (DC) power supply is used, it is referred to as pulsed-ECH or PECH. Consequently, ECH and PECH combine the capabilities and advantages of their constituent processes while at the same time limiting their individual disadvantages.

The most important capabilities and advantages of ECF are (1) surfaces finished by ECF or PECF is largely free from mechanical damage including residual stresses, cracks, microcracks, hardness alternations, plastic deformation, and free from thermally induced damage such as recrystallization, microstructural changes, and heat-affected zones; (2) high MRR when compared to some of the other advanced finishing processes; (3) dependence of process performance on the mechanical properties of the workpiece material which implies that it can finish material of any hardness; (4) negligible tool wear largely due to the noncontact nature of the process.

The main capabilities and advantages of mechanical honing include (1) ability to correct errors/tolerances related to geometry or shape of the workpiece; and (2) controlled generation of tribologically important surfaces with a cross-hatch lay pattern that improves lubricant retention.

The most significant limitation of the ECF/PECF processes is the formation of an inactive or passivation layer on the anodic workpiece in the form of a metallic oxide during electrolysis that prohibits further electrolytic dissolution of the workpiece. The protective layer is generally more pronounced (thickness) in the valley regions of the surface topography.

The most significant limitations of the mechanical honing are (1) possible mechanical damage to the workpiece; (2) unpredictable failure of the honing tool; (3) reduced honing tool life; and (4) reduced productivity due to the process being slow with interruptions caused due to frequent tool failures and changes.

Hybridization of ECF/PECF and mechanical honing may therefore overcome their individual inherent limitations while exploiting their advantages. This usually implies that the majority of workpiece material (i.e., 80%–90%) is removed by ECF/PECF during the hybridized process and that the role of honing is restricted to (1) removal of the prohibiting oxide passivation layer; and (2) selectively removing surface peaks that are the result of nonuniform material removal by the electrolytic dissolution process. This may consequently improve the surface characteristics and quality of the gears significantly, thereby making ECH/PECH a preferred alternative that is economical and sustainable.

The concept of the ECH process evolved in 1960s. Finishing by ECH involves maintaining a required interelectrode gap (IEG) in the range of 0.1–1.0 mm between the anodic workpiece and the cathodic tool subject to a DC voltage in the range of 8–30 V. The IEG is further flooded with an appropriate electrolyte. The electrolyte facilitates an uncontaminated environment in the IEG by flushing away products of the electrolytic dissolution. The electrolyte flow is also important for heat removal due to the heat generation of the DC current and the electrochemical reactions. In the PECH process, a pulsed DC power source is used to provide short voltage pulses between the anode and cathode so that electrolytic dissolution of the workpiece selectively occurs during the pulse-on time only. Pulse-on and off times typically range between 1 and 7 ms, and 2 and 15 ms respectively. Effective flushing occurs during the pulse-off time. Successive application of the honing tool then selectively removes the metallic oxide passivating layer enabling more effective electrolytic dissolution.

Until the early 1980s, the use of ECH was limited to finishing internal cylinders, bushes, and bearings. Capello and Bertoglio [14] used it for the first time to finish *hardened helical* gears of 2.5 module (17 teeth) by a specially designed cathode gear with 64 teeth on an apparatus with both reciprocating and rotary motion of the workpiece and cathode gear while controlling the IEG. Their initial results were not satisfactory for helix and involute profiles, but it did confirm the possibility of using ECH for gear finishing. Their results did however highlight the need for designing the cathode gear to take into account the electrochemical process control parameters. Chen et al. [15] used ECH for finishing of *spur* gears by using the arrangement as depicted in Fig. 5.7A and reported improvement in the surface roughness of the flank surfaces and profile accuracy of the spur gears. It also resulted in a noise level reduction of between 5 and 8 db. Wei et al. [18] varied the electric field intensity to control the electrolytic dissolution seamlessly along the *spur* gear tooth profile thereby improving its accuracy in a process referred to as field-controlled ECH, whereas He et al. [19] used the time-control method to correct profile errors in *spur* gears in a process which they referred to as slow scanning field-controlled ECH. They reported on the perceived superiority of the time control method over the current field control method.

The design of the cathode and honing gear and their arrangement with respect to the workpiece gear in the ECH finishing chamber depends largely on the geometry of the workpiece gear. The experimental setup consists of the following typical subsystems:

- **DC power supply system**: DC 0–100 V, current 10–110 A with programmable options for controlling pulse-on and off duration and timing, operable as either a constant current source or as a constant voltage source is used. The positive terminal is electrically connected to the workpiece gear, whereas its negative terminal is connected to the cathode gears via its mounting shaft through a carbon brush and slip-ring assembly.
- **Electrolyte supply and cleaning system**: Electrolyte is essential for electrolytic dissolution. Additionally, it removes the generated process heat and the solid process products from IEG. Desirable characteristics of the electrolyte are high electrical conductivity, low viscosity, high specific heat, increased chemical stability, resistance to formation of a passivating layer on the workpiece, noncorrosive, nontoxic, economical, and readily available. Generally, an aqueous solution of salt-based electrolytes such

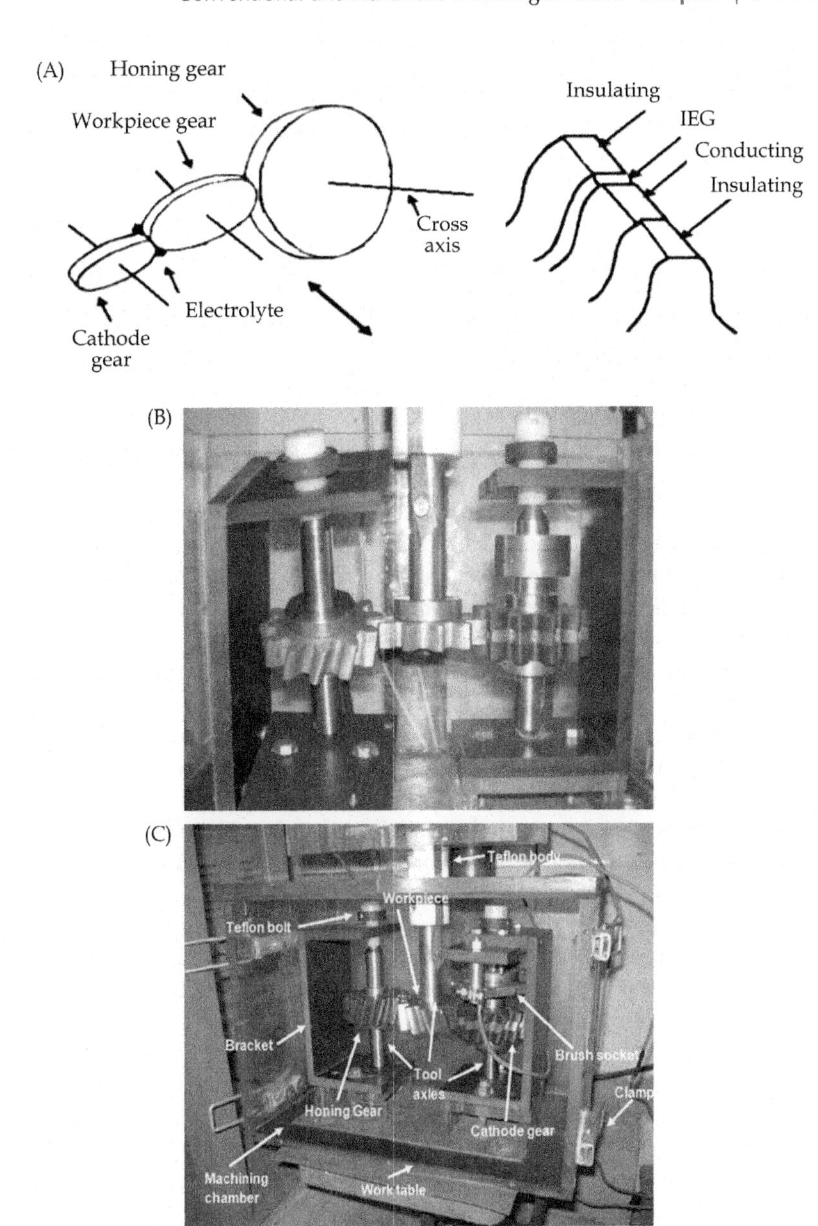

FIGURE 5.7 Finishing of cylindrical gears by ECH: (A) arrangement of workpiece gear, cathode gear, and honing gears; (B) finishing chamber developed by Naik [16] for spur gear finishing by ECH; and (C) finishing chamber developed by Mishra [17] for helical gears.

as sodium chloride (NaCl), sodium nitrate ($NaNO_3$), sodium chlorate ($NaClO_3$), or their combinations are passed through the IEG at a flow rate of 10–50 L/min. The electrolyte supply and the associated cleaning system is designed to deliver the filtered electrolyte at the required flow rate, pressure, and temperature to the finishing chamber and then to subsequently recirculate it back to the storage tank. It consists of a stainless steel pump capable of delivering a wide range of pressures and flow rates, filters, piping, control valves, heating, and cooling coils to maintain a constant electrolyte temperature, pressure, and flow rate measuring devices and storage and settling tanks. Generally, an electrolyte tank of storage capacity of 200–400 L is used. The electrolyte is filtered by using relatively coarse filters manufactured from anticorrosive materials such as stainless steel and Monel. Separation by centrifuge or magnetic field may also be utilized. Corrosion is controlled by manufacturing the piping and tanks from materials such as stainless steel, glass-fiber-reinforced plastics, plastic-lined mild steel, or concrete.

- **Supporting machine tool**: The finishing chamber needs to be supported while providing variable speed rotary motion to the workpiece gear. Typically, this is facilitated by DC motor with a speed range of 30–1500 rpm. For finishing cylindrical gears, reciprocating motion of the workpiece gear is required. A stepper motor driven kinematic assembly is usually used. The finishing chamber is mounted onto the machine tool base with appropriate electrically insulating arrangements. Machine tool design considerations include corrosion resistance and adequate power for the machining torque without deforming the workpiece. A sturdy bench (vertical) drilling machine or vertical milling machine is the best suited for this purpose.
- **Finishing chamber**: The cathode, workpiece, and honing gears are all arranged and mounted in the finishing chamber. A transparent polymer such as Perpex is typically used to fabricate the finishing chamber due to its adequate strength and being able to facilitate improved visualization of the ECH/PECH process. Pedestal ball bearings are used to mount and support the stainless shafts on which the honing and two cathode gears are mounted. Bakelite and/or stainless steel blocks are typically used in the finishing chamber to retain the various support bearings due to their corrosion resistance and adequate strength. The following sections describe in more detail the finishing chambers, setup, and operating principles of ECH for cylindrical and conical gears.

5.2.1.1 ECH of Cylindrical Gears

Naik [16] and Mishra [17] developed dedicated finishing chambers for ECH of spur and helical gears. These are illustrated in Fig. 5.7B and C.

In both finishing chambers, the workpiece gear meshes with the cathode gear on one side where material removal by ECF occurs, while on the other side, mechanical honing is facilitated to selectively remove the metallic oxide passivation layer from its flank surfaces. The cathode gear was specially designed by sandwiching an electrically conducting and undercut layer between two nonconducting layers which mesh with the workpiece gear. This design avoids short-circuiting between the cathode and workpiece gears while simultaneously maintaining the appropriate IEG. The axes of the cathode and workpiece gears are parallel. For finishing of spur gears, the honing gear is mounted on a floating arrangement in a cross-axis layout (Fig. 5.7B) with the workpiece spur gear to ensure dual flank contact. Finishing of helical gears by ECH do not require such an arrangement because the same objective is achieved by utilizing a cathode and honing gear with an opposite helix angle than that of the workpiece gear (Fig. 5.7C). For a right-handed helix workpiece gear, the cathode and honing gears are arranged with a left-handed helix angle and vice-versa. The honing gear may be either an abrasive impregnated gear or made from a material with a higher hardness than that of the workpiece gear material. The workpiece gear is driven, whereas the cathode and honing gears rotate due to appropriately close meshing with the workpiece gear. The workpiece gear is reciprocated along its main axis to ensure finishing of the entire face width while maintaining the appropriate IEG state (electrolyte flow and DC voltage). All three gears have the same involute profile.

5.2.1.2 ECH of Conical Gears

Finishing of conical gears by ECH or PECH is more challenging than cylindrical gears due to their complex geometry that prevents reciprocation of the workpiece gear which is required for finishing the full face width. To address this challenge Shaikh [20] conceptualizing a novel idea of using twin-complementary cathode gears as shown in Fig. 5.8A along with the development of an appropriate finishing chamber based on the concept for precision finishing of conical gears by ECH. In this concept, one of the complementary cathode gears ("3" in Fig. 5.8A) has an insulating layer sandwiched between two conducting layers whereas, the other complimentary cathode gear "4" has a conducting layer sandwiched between two insulating layers. The conducting layer is undercut by 1 mm as compared to the insulating layers to maintain the required IEG between the cathode and anode gears. The workpiece gear "1," honing gear "2," and complementary cathode gears "3" and "4" are mounted on their respective shafts in such a way that their axes of rotation are coplanar and perpendicular to each other. Only the workpiece gear is driven. The other gears rotate due to an appropriately close meshing with the workpiece gear. Jets of electrolyte "5" are supplied to the IEG along with correct DC voltage. The finishing of the workpiece gear by electrolytic dissolution then occurs due to the two complementary cathode gears while the passivating oxide layer is simultaneously removed by the honing gear on a different part of the gear. This enhances productivity while maintaining the appropriate quality when finishing conical gears. Pathak [21]

(A)

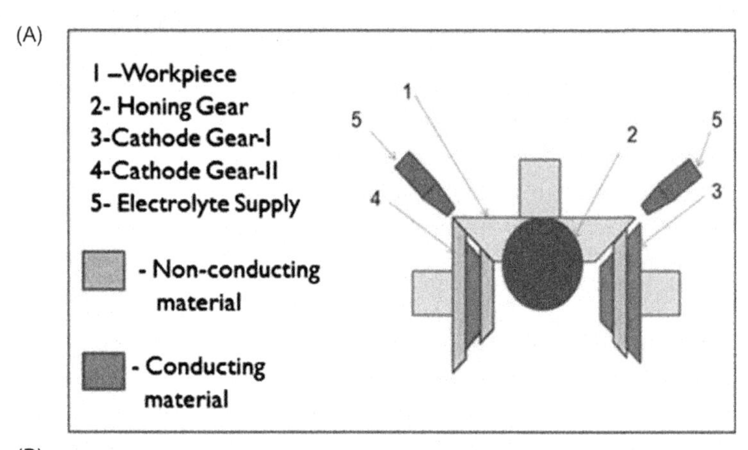

(B)

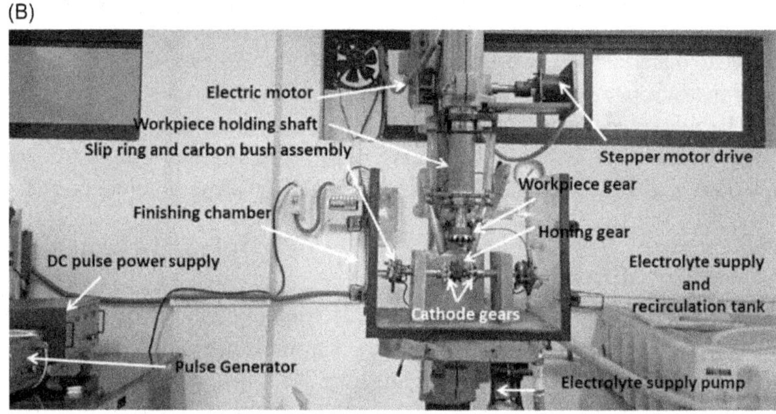

FIGURE 5.8 Finishing of conical gears by ECH: (A) concept of twin-complementary cathode gears proposed by Shaikh [20] and (B) setup and finishing chamber by Pathak [21].

further improved the finishing chamber developed by Shaikh [20] by automating the movement of the workpiece gear which enabled mounting of a number of workpiece gears simultaneously which reduces setup time. The deflection of the complementary cathode gears and honing gear support shafts were also significantly reduced by using stainless steel blocks for support. Fig. 5.8B depicts the setup with the finishing chamber improvised by Pathak [21] for finishing bevel gears by ECH and PECH.

The following are important process parameters along with their variation ranges as appropriate to ECH and PECH:

- **DC-power-supply-related parameters** such as voltage (6−30 V); current (100−3000 A); current density (12−47 A/mm^2); pulse-on time (1−7 ms); pulse-off time (2−15 ms); duty cycle (10%−66%).

- **Electrolyte-related parameters** such as type (i.e., acidic, alkaline, or salt based), composition, concentration, temperature (generally up to 38°C), pressure (0.5–1.0 MPa); flow rate (10–50 L/min); electrical conductivity; specific heat; viscosity; boiling point.
- **Workpiece-related parameters** such as chemical composition; electrochemical characteristics (valency of electrolytic dissolution dissolution); atomic weight; density; electrical conductivity; specific heat.
- **ECF-related parameters**: continuous or pulsed; IEG (0.1–1 mm); speed of rotation (20–100 rpm); speed of reciprocation (0–18 m/min).
- **Honing-related parameters** such as honing tool design; pressure exerted by honing tool (0.5–3 MPa); type of honing tool (abrasive impregnated or harder than workpiece material); hardness; abrasive type (i.e., Al_2O_3, SiC, CBN, diamond); abrasive size (70–1200 mesh size).

Shaikh [20] achieved surface finish improvements of up to average roughness (R_a)—49.4% and maximum roughness (R_{max})—42.7% when finishing straight bevel gears by ECH. Pathak [21] achieved DIN 6 gear quality and improved surface finish up to 47% by finishing by PECH.

5.2.2 Gear Finishing by Electrochemical Grinding

Gear finishing by ECH reduces to gear finishing by electrochemical grinding (ECG) (1) if the honing gear is replaced with a grinding wheel; or (2) role of mechanical honing process becomes insignificant. The basic principles as applicable to ECG are therefore similar to those previously introduced for ECH. MMRs in the ECG can be increased by (1) introducing suitable abrasives into the cathode gear; or (2) mixing abrasive particles into the electrolyte. Abrasives such as aluminum oxide, SiC, or CBN can be used depending upon the hardness of the workpiece gear material. ECG is a particularly unique and useful solution for selected applications that include reprofiling of worn locomotive traction motor gears [22].

5.2.3 Gear Finishing by Abrasive Flow Finishing (AFF)

Abrasive Flow Finishing (AFF) is an advanced process for precision finishing of gears. It was developed in the 1960s to achieve a mirror-like surface finish at minimum cost. This process is used to deburr, polish, and increase edge radiuses in difficult-to-access regions of complex parts by flowing an abrasive laden self-deformable semisolid fluid, known as the "AFF medium", through and/or over them. The displacement of the AFF medium typically occur due to extrusion pressures of up to 220 bar and with flow rates up to 400 L/min. The abrasives in the fluid shear off surface peaks imparting a nanofinish and mirror-like appearance. More aggressive finishing takes place at those regions where localized restriction to the flow increases the localized pressure or speed. A typical AFF machine consists of a hydraulic cylinder(s) and piston(s), AFF medium, fixture or tooling for the gear to be finished, limit switches, supporting structure, pressure control valve, stroke counter, and a hydraulic power unit (see Fig. 5.9). Modern AFF machines are equipped with a CNC control system to facilitate mass production.

The AFF medium is a mixture of appropriate abrasive particles, pliable and viscoelastic polymer, and a suitable lubricating oil. It is forced to flow along a path formed by the tooling/fixture and/or workpiece. Commonly used abrasives are Al_2O_3, SiC, CBN, and diamond. Their size ranges from 0.005 to 1.5 μm. The viscosity and flow rate of the AFF medium significantly affect MMR and edge-radius size. The viscosity of the AFF medium can be controlled by changing the ratio of the polymer and lubricating oil. The flow rate of the AFF medium depends on the size of the passage and hydraulic pressure. Low viscosities and higher flow rates are used for deburring and radiusing sharp corners, whereas low and steady flow rates are used in finishing and polishing applications [22].

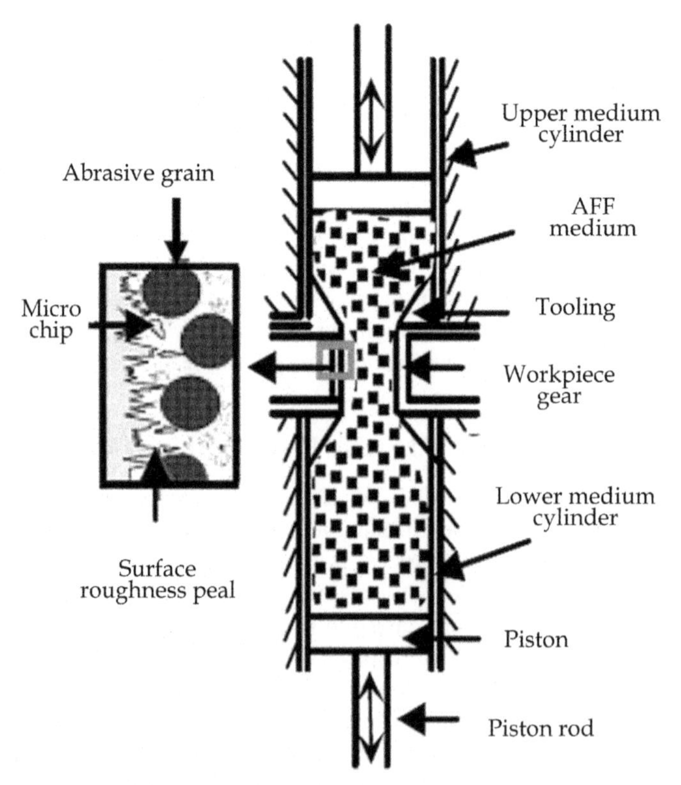

FIGURE 5.9 Working principle of a two-way AFF process [23].

The AFF process may be performed in three ways depending on the flow path and/or motion of the AFF medium:

- One-way AFF: AFF medium flows in one direction only in a cyclic loop [24].
- Two-way AFF: Most promising and extensively used AFF approach, and generally used for commercial gear finishing applications. In this process, the AFF medium is moved to and fro between two pistons each contained in a cylinder directly opposed to one another and either side of the workpiece passage and/or tooling (see Fig. 5.9). The setup may be arranged either vertically or horizontally. One or the other cylinder is filled with the AFF medium before it is then flowed by an appropriate piston stroke at high pressure toward the opposing cylinder. The process then repeats itself from the other side. One complete cycle consists of a single stroke from each side. The abrasive particles contained in the AFF medium then effectively shears off and abrades the surface peaks from the workpiece surfaces thus imparting a precision finish. The process can either be operated for a fixed number of cycles or until the desired surface finish is achieved [25].
- **Orbital AFF**: This version of the AFF process is similar to two-way AFF except that finishing of the workpiece is assisted by imparting high frequency but low amplitude vibrations to the workpiece in the direction normal to the movement of the AFF medium [26].

Recent investigations [27–29] have confirmed that the AFF process has the potential to be a productive, economical, and sustainable alternative finishing process for gears but specifically for those that are manufactured from strong and/or hardened materials. It can also improve the surface finish and microgeometry of gears in comparatively less time. Kenda et al. [27] compared the microgeometry of a gear manufactured by wire electric discharge machining (WEDM) and subsequently finished by the AFF process. They reported that the required cambered shape of the gear tooth could be largely facilitated by the AFF process. Xu et al. [28] used AFF to finish helical gears and reported that it increased tooth stiffness and load carrying capacity while reducing the contact area, friction, and wear of the flank surfaces. Venkatesh et al. [29] used US-AFF for finishing straight bevel gears and reported improvement in its surface finish. Further research on the continued and improved use of AFF of cylindrical and conical gears is ongoing.

The important process parameters that affect the performance of the AFF process are as follows:

- **AFF medium-related parameters**: Type, size, and concentration of abrasive particles; type and concentration of the polymer carrier; additives; type of lubricating oil; rheological properties of AFF medium.
- **Workpiece-related parameters**: Nature (i.e., ductile, brittle); hardness; toughness; geometry; initial surface roughness; initial surface integrity.

- **Machine-related parameters**: Extrusion pressure; number of cycles; provision to monitoring of the finishing process and ability to control rheological properties of the AFF medium; in-situ measurement of surface roughness.

The following are the salient features of the AFF process [22–29]:

- AFF has the ability to finish inaccessible or difficulty-to-access areas on complicated components including various gear types of typical geometry.
- AFF may produce average surface roughness values up to 50 nm; achievable dimensional tolerances up to ± 5 μm with good repeatability.
- It can finish multiple gears simultaneously in minimum time with uniform, repeatable, and predictable results. Automated AFF system can finish thousands of gears per day thus reducing labor cost significantly.
- It offers flexibility to accommodate changes in material, geometry, and dimension of gears, AFF medium, abrasives, and AFF process parameters. This makes it possible to finish a variety of gear types by using the same setup or by selective modification to the fixtures/tooling for holding purposes.
- It is particularly suitable for finishing strong and/or hardened materials such as cast iron; super alloys; nickel alloys, titanium-based alloys, ceramics, and carbides.
- The risk of surface integrity-related effects including residual stress, metallurgical transformations, plastic deformation, tearing, and cracking are significantly reduced.

5.2.4 Other Advanced Finishing Processes for Gears

This section introduces selected gear deburring and superfinishing processes. A burr is a small 3D feature projecting from a manufactured and finished component. Burrs are individually classified as either compressive, corner, edge, entrance, exit, feather, flash, hanging, parting-off, roll over and tear [30,31]. Its presence on flank surfaces of manufactured gears adversely affects its wear characteristics, service life, operating performance, and aesthetics. Removal of burrs is also important to avoid injuries to the operators involved in lubrication and maintenance of the gears. The presence of burrs can be limited by accurate control of the gear-manufacturing process but complete removal may be required in view of the tribological importance of the flank surfaces of a gear and the adverse effects to these surfaces due to the presence of significant burrs. Different types of gear deburring processes, such as mechanical, thermal, and electrolysis-based are discussed in the subsequent sections.

5.2.4.1 Water-Jet Deburring

The working principle of WJD is similar to that of water-jet machining (WJM) except that WJD is conducted with lower water pressures, typically between 30 and 50 MPa (up to approximately 100 MPa in exceptional cases) as compared to WJM that may go up to 400 MPa. Highly pressurized water exiting through a nozzle gains significant kinetic energy by virtue of the high fluid velocity. This high-velocity water jet is directed toward the target surface in such way that it selectively removes those materials which are either not solidly attached to the gear surface or is an undesired product of the machining and/or finishing process. It facilitates the removal of chips, debris, loosely attached metal shavings, burrs, and feather-edge burrs (which are visible by microscope only) from the flank surfaces of a gear while simultaneously cleaning them without affecting the basic gear geometry and material characteristics. Fig. 5.10 illustrates WJD of an external spur gear.

Sometimes burrs cannot be removed by WJD, but it is still used as a high-pressure water jet which is highly effective for cleaning the part. A simple test, mimicking the pressure exerted by a 0.9-mm water jet operating at 48 MPa, may be used by attempting to remove the burr with a 0.5-mm lead pencil. If the burr can be deformed or removed with the pencil chances are that it can be removed by WJD [32]. If a gear burr cannot be removed by WJD alone, it may be combined with mechanical deburring using a rotary brush or filament brush. The gear is first mechanically deburred and then cleaned by WJD to prepare for use in critical assemblies.

A WJD machine typically contains the following subsystems:

- **Hydraulic pumping unit**: Its function is to generate a high fluid pressure. It consists of an electric motor that drives a pump that is essentially the heart of a WJD system. Generally, positive displacement type plunger pumps are used due to their ability to generate high constant water pressure. The pump power depends upon the required water flow rate to adequately supply to a bank of nozzles. The greater the water flow rate for a

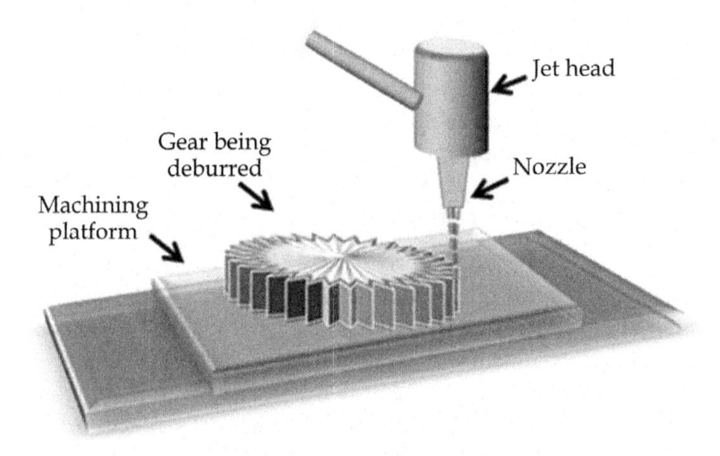

FIGURE 5.10 Water-jet deburring of an external spur gear.

given pressure, the higher the required pump power. One or more high-pressure shift valves may be used to direct water from the pump to the deburring station.

- **Tooling unit**: Consists of a nozzle to produce a high-velocity water jet and workpiece table to hold the workpiece with respect to the nozzle and to provide the required relative motion to complete the deburring process. Correct selection of WJD tooling is important to minimize deburring time. Commonly used materials for the nozzle are HSS, carbide, sapphire, ceramics, and some exotic materials. Harder material results in longer nozzle life. Nozzle diameters typically range from 0.05 to 0.5 mm but may be as large as 1 mm with a typical expected life of 250–500 hours. Nozzles may fail due to abrasion caused by dirt particles in the water or constriction due to water mineral deposit build up. To avoid this, water may be treated with softeners or deionizers and filtered (down to 1 μm filter elements) to minimize damage to the nozzle. A "direct" nozzle is used to produce a solid stream or jet of fluid that is selectively directed toward the burr to be removed.

  Rotary fan nozzles are used for entering small diameter bores down to 6 mm or small cavities. Generally, three or more fan nozzles are rotated as the workpiece is translated through the bore for deburring. For high volume applications or when a reduction in deburring time is required a custom manifold may be designed that deburr all features simultaneously [32].
- **Catcher unit**: It is used to reduce the associated noise of the water-jet process and to minimize the exposed length of the water jet for safety purposes. The noise intensity level is a function of the water pressure, i.e., jet velocity. Due to the relative violent nature of the process, mist, spray, and removed burrs may be ejected into the immediate environment of the machine tool that may affect health and safety concerns. Typically a tube or tank is used to collect the water. Generally a tube with a length of 300–600 mm is positioned beneath the water jet and connected to a drain hose. A short tube with a hard replaceable insert, to diffuse the water jet quickly, can also be used. The recovered water is strained and filtered and recycled back to a clean water tank where it may again be filtered before it is once again fed to the high-pressure pump. Usually, a significant portion of the pump power is dissipated as heat to the water. This may necessitate a heat exchanger or water chiller to control the water temperature.

The time required for effective WJD is a function of the machine tool type, pump power, sophistication of the nozzle tooling, and most importantly the number of features that are to be deburred. Typically, it takes 5–10 seconds to deburr per part feature while total deburring time may be anything between 30 and 60 seconds. WJD has some unique advantages when compared to other gear deburring processes. These include (1) gear is fully cleaned, burr-free, residue-free, and ready for assembly and use; (2) it is environment-friendly because it does not use any abrasives or corrosive chemicals and occurs at room temperature. The most suitable gear materials for effective WJD are aluminum, brass, cast iron, alloy steel, high strength steels, polymers, and composites. Harder materials require higher water pressures and sharp edges will be mostly retained. WJD has a wide range of application that may include automotive, aerospace, biomedical, MEMS, fluid power, food processing, chemicals, home appliance, etc. It is especially useful for (1) deburring inaccessible features of components; (2) deburring those components and materials for which thermal-based or corrosive-chemical-based deburring processes cannot be used; (3) edge deburring and cleaning of gears to remove loosely attached metal shavings that may require additional honing and/or broached keyways; (4) deburring and washing to remove particles that cannot be removed with conventional cleaning processes; (5) where consistency in quality of deburring is needed; and (6) delivering parts cleaned all the way down to the microscopic level to the point of assembly.

5.2.4.2 *Electrolytic Deburring*

Electrolyte deburring (ED) is an electrolysis-based process to remove burrs from inaccessible or difficult-to-access locations on a manufactured component. It has evolved as an attractive and legitimate alternative for deburring of critical and tribological sensitive components such as gears, bearings, bushes, etc. Flank surfaces of gear teeth are fine finished by effective burr removal. Applicable burr sizes range from 0.02 to 0.3 mm but exact dimensions of burr and sharp corners that may be finished depend on the application and design [33].

In this process, a preshaped cathode is fixed at a predetermined distance from the gear edge such that the burrs are located in the path of the current flow without touching the cathode as shown in Fig. 5.11. A short-circuit detection system is used to ensure that no burrs touches the cathode before the DC power is applied. The voltage and amperage setting is a function of the burr size and finishing requirements. The cycle time may then be estimated based on the selected process parameters. Deburring commences with the application of the potential difference and subsequent current flow at the workpiece gear. Deburring occurs in a selective manner by arranging the process such that the areas to be finished are preferentially found in high current density areas. This limits dissolution of the stock material. Modern ED machine tools are designed with the ability to fully control (usually PLC based) all applicable process parameters to ensure repeatable part quality. The duration of an ED process, associated with a gear, is a function of the burr size and surface finish requirements, and may therefore range from as little as 5 seconds to some minutes.

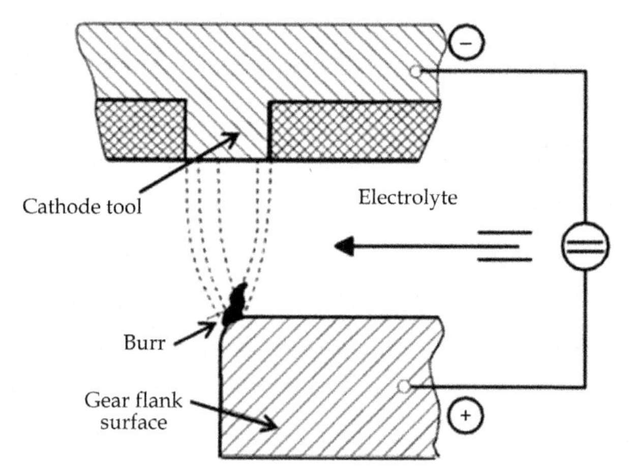

FIGURE 5.11 Schematic of electrolytic deburring (ED).

ED of gears offers certain distinct advantages over other deburring processes. These include stress relieving of the gear surface; removal of impurities embedded in the gear during its manufacturing; improving surface finish and enhancing overall appearance; may extend service life of the gears; and requires minimum operator training. There are however certain environmental and safety concerns which should be addressed when using this process. The burrs are electrochemically dissolved into the typically salt electrolytes forming a sludge that needs to be replaced and properly disposed of according to environmentally friendly practices and local legislation every couple of months. This sludge may include heavy metals depending on the materials involved. Generally, it is not suitable for job-shop production and for every manufacturer.

5.2.4.3 Deburring by Thermal Energy Method

The thermal energy method (TEM) process deburrs components by igniting a volatile gas mixture within a pressurized chamber containing the work piece gear. This creates a combustion heat wave that almost instantaneously reaches temperatures up to 3500°C that is then subsequently only sustained for a few milliseconds. The wave passes over the component and burns off the burrs and flashing. Essentially, the burrs and flashings autoignites in an oxidizing reaction because of their small relative size. The rest of the component acts as an effective heat sink. Because of the enveloping nature of the gas mixture, burr and flashing removal is effective and complete regardless of position.

Fig. 5.12 presents a graphic illustration of the TEM process. It does not require special tools and typically uses a mixture of hydrogen and oxygen in a ratio of 0.66–2 and a spark plug to ignite the mixture. Follow-on heat

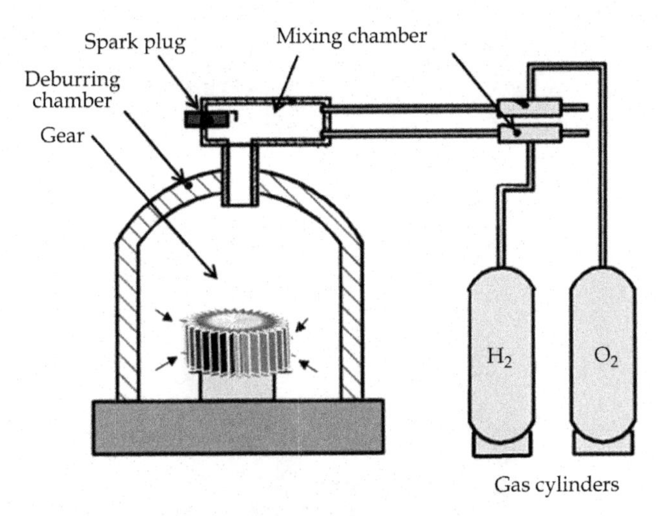

FIGURE 5.12 Thermal energy method (TEM) deburring.

treatment or immersive cleaning processes may be required to remove the loosely attached oxide surface powder layer that results from the process. The process is fast and is usually complete within 20 seconds.

The performance of the TEM process depends on (1) size and position of the burr; (2) volume-to-surface ratio of the burr; (3) melting temperature, ignition temperature, and heat conductivity of the gear material; and (4) composition and pressure of the dielectric gases used which depends on the gear material. Gas pressure in the range from 10 to 60 bar is used for the gears made from various materials.

5.2.4.4 *Brush Deburring*

Brush deburring involves removal of burrs and flashings on flank surfaces of gear teeth by subjecting them to high-speed filament impact. Filaments mounted as a rotating brush impart their kinetic energy via impact to the burr or sharp edge. Together with the built in compliance associated with filament kinematics localized undesirable edge effects may be removed or modified without altering the overall part geometry. Brushes may be classified according to their geometry and the type of filament. Brush geometry is closely linked to the workpiece geometry that is to be deburred and access to the area of interest. These may include wheel, cup and end type brushes (see Fig 5.13A). Typical filament materials include steel, natural fibers, and synthetic fibers. Steel filament brushes usually come in two types. The crimped type uses filaments where each strand of wire is crimped or bent, whereas for the straight type, the wire is used as is but is either knotted or twisted to retain it on the wheel circumference. These wire brushes essentially work by an impact-driven mechanism and therefore requires elevated speeds with

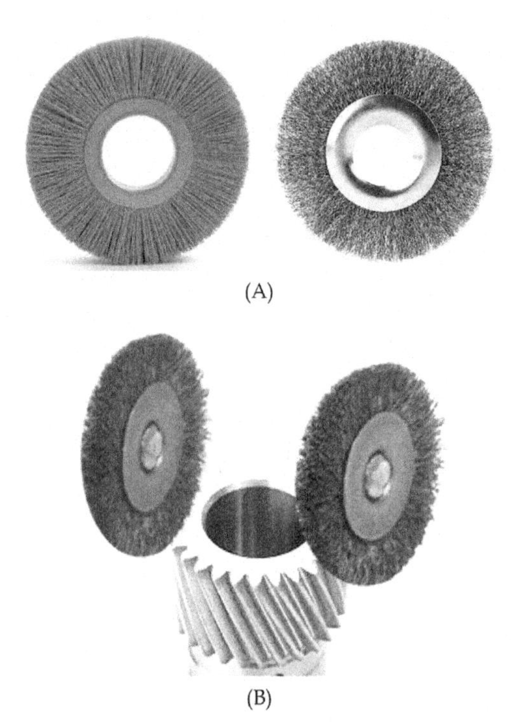

FIGURE 5.13 (A) brushes used for gear deburring and (B) brush deburring of a gear.

minimal penetration. The crimped type filaments are less aggressive and preferentially used for applications where improved control and smaller defects are encountered. The straight type wire filament brushes are preferentially used for removal of large and heavy burrs.

Filaments manufactured from synthetic fibers such as nylon impregnated with abrasive particles are also extensively used. These rely more on the filament being dragged, causing abrasion, over the surface or defect for removal than pure impact. They therefore require lower speeds and greater penetration. Being not as aggressive as wire filaments, they are best used on ground gears and gears manufactured from exotic materials.

Natural fiber filaments are mostly used for deburring of gears manufactured from exotic materials and buffing and polishing.

Although brush deburring seems to be a relatively simple solution there are many factors that need to be considered and affect its performance. These are (1) type, material, and direction of brush; (2) speed and feed of brush; (3) positioning of the brush with respect to the gear; (4) duration of brush engagement with the gear and its control; and (5) selection of coolant. The technology associated with brush deburring has improved considerably. This has resulted in significant advantages such as (1) rugged machine tool frames and mechanical features; (2) automatic compensation for wear of unattended brushes using actual loads; (3) smaller and safer enclosed machine tools; (4) large access opening for setup, brush change, and maintenance; (5) dust collectors integrated into the machine; (6) coolant contained within the machine enclosures; (7) automatically controlled machines that use variable-frequency drives which allows the brush to be rotated at the optimum speed; and (8) more efficient use of electrical energy [34].

Fig. 5.13B presents a typical optimal brush setup for deburring of gears. It removes burrs efficiently and maximizes the brush life.

Recent developments in this field has led to the concept of *zero-setup gear deburring*, which uses four steel wire brush wheels simultaneously to quickly deburr the gears in a single operation thus eliminating many variables and making use of the most appropriate positioning of the brushes [34].

5.2.4.5 Chemically Accelerated Vibratory Surface Finishing

Chemically accelerated vibratory surface finishing (CAVSF), also referred to as isotropic surface finishing or superfinishing, is an inexpensive and environmentally friendly process capable of reducing stress raisers, peaks, and the damaged material from the flank surfaces of a gear. This process can decrease the average surface roughness of a case hardened gear from 0.3 to 0.05 μm while only removing approximately 5 μm material from the surface and thereby producing minimal geometry change of the gear [35].

The CAVSF of gears occurs in a vibratory finishing bowl or tub using a high-density, nonabrasive ceramic media, and a continuous flow of a mildly acidic (nominal pH value approximately 5.5), nontoxic, and nonhazardous chemical solution (see Fig. 5.14). The media in this case refers to the simple discrete ceramic elements within which the gears to be finished is placed. These are nonabrasive because it does not contain discrete abrasive particles, i.e., it is unable to abrade material from the hardened surface of the gears being finished on its own and are usually smooth. The size and shape of the media is chosen such that it works best with the gears to be finished. The vibratory motors typically run at a speed of 1800–2000 rpm. The chemical solution forms a stable and soft coating on the surfaces of the gears. The rubbing motion produced by the movement of the ceramic media then effectively wipes this "conversion coating" from the peaks on the gear surfaces and thereby exposing the same peaks again to "fresh" solution. The valleys are largely unaffected because the solution is not mechanically removed in the same manner as for the peaks. No finishing occurs where the ceramic medium is unable to contact and thereby remove the solution. The solution (conversion coating) is continuously formed and scrubbed off imparting a surface smoothing mechanism. This process continues until the required surface finish is attained. The finished gears are then initially rinsed in tap water followed by demineralized water to wash away the chemical solution. The gears may then be treated with a rust inhibiter.

FIGURE 5.14 Vibratory bowl with ceramic media used in chemically accelerated vibratory superfinishing of gears.

This process yields a surface texture that is conducive to lubrication, free of stress raisers, damaged material, and peak asperities. Gears finished by CAVSF typically exhibit lower friction, lower operating temperature, less wear, better scuffing resistance, and higher contact fatigue resistance.

5.2.4.6 Black Oxide Finishing

Black oxide finishing of gears involves the formation of a thin (approximately 0.5 μm) but stable marginally corrosion resistant and durable oxide layer on the surface. This layer usually has a black appearance henceforth the name. It is porous and therefore has benefits as far as the retention of rust inhibitors or lubricants are concerned. It increases corrosion resistance in storage and service, increased durability for gears and power transmission components without affecting their dimensions, assembly and operation [36]. Fig. 5.15 shows an external spur gear finished by black oxide finishing. It is a nearly ideal process for finishing those precision manufactured components that cannot tolerate the variable thickness of paint or electroplating. Three different black oxide finishing processes are generally used. They are differentiated relative to the heat requirements and chemicals used in the process. All three blackening processes provide dimensional uniformity and stability.

- *Conventional black oxidizing* occurs in a boiling caustic soda (NaOH) bath at approximately 160°C. This produces a black iron magnetite finish

FIGURE 5.15 Attractive black oxide finish on an external spur gear.

in 20–30 minutes. The overall finishing quality is good except for cast iron gears on which a red coating is formed and on parts with blind holes or recesses in which white salt leaching occurs.

- *Room temperature blackening* uses copper/selenium without the addition of any heat to produce a black conversion coating in 2–5 minutes. Although it may be used as a substitute to conventional black oxidizing in many applications, the finish is less durable due to the use of copper and selenium. It also requires the use of an ion exchange add-on system to purify and recycle the rinsing water before it may be safely disposed of.
- *Low-temperature black oxide* finishing was more recently developed and implies the formation of a durable black magnetite finish produced within a relatively short time (± 10 minutes) by using 80% less caustic soda than conventional black oxide finishing. This eliminates the chemical hazards associated with the conventional black oxide finishing process thus making it easier and safer to operate and use. Red coating and white salt leaching problems that are often observed with the conventional process is largely avoided. The rinse water is also nonhazardous and can be discharged to sewerage system without waste treatment.

The basic process involves the following steps:

- **Cleaning** of the gear by water-soluble solutions thereby removing the machining fluid, coolants, and other foreign particles.
- **Rinsing** of the cleaned surface by tap water for 30 seconds.
- **Surface conditioning** by a mild acidic surface conditioner to remove minor oxides and to deposit a primer coating that facilitates a uniform final black oxide coating.
- **Rinsing** to remove residue, if any, of the surface conditioner.
- **Blackening** by forming the oxide coating in approximately 20–30 minutes at 145°C (conventional process) or in approximately 10 minutes at 95°C (low-temperature process).

- **Rinsing**: A final rinse to remove any black oxide residue from the gear surfaces. If the high temperature process is used, then the rinsing water may have a higher pH value therefore it must be neutralized before disposal.
- **Sealing** makes use of the rust inhibitors (water-based or solvent-based) considering the intended end use of the gear.

REFERENCES

[1] B. Karpuschewski, H.J. Knoche, M. Hipke, Gear finishing by abrasive processes, Ann. CIRP 57 (2) (2008) 621–640.

[2] N.K. Jain, A.C. Petare, Review of gear finishing processes, in: M.S.J. Hashmi (Ed.), Comprehensive Materials Finishing, 1, Elsevier, Oxford, 2016, pp. 93–120. Available from: http://dx.doi.org/10.1016/B978-0-12-803581-8.09150-5.

[3] I. Moriwaki, T. Okamoto, M. Fujita, T. Yanagimoto, Numerical analysis of tooth forms of shaved gear, JSME Int. J. Ser. III 33 (4) (1990) 608–613.

[4] F. Klocke, T. Schroder, Gear shaving: simulation and technological studies, in: International Design Engineering Technical Conferences and Computers and Information in Engineering Conference, ASME, 2003, pp. 257–264.

[5] ⟨www.mhi-global.com/products/detail/ind_gear_shaving_cutter.html⟩.

[6] J.S. Zhong, L.S. Qu, Intelligent control of the gear-shaving process, Proc. IMechE, B: J. Eng. Manuf. 207 (3) (1993) 159–165.

[7] R.H. Hsu, Z.H. Fong, Theoretical and practical investigations regarding the influence of the serration's geometry and position on the tooth surface roughness by shaving with plunge gear cutter, Proc. IMechE, C: J. Mech. Eng. Sci. 220 (2006) 1170–1187.

[8] S.B. Rao, Grinding of spur and helical gears, Gear Technol. 9 (4) (1992) 20–31.

[9] H.K. Tonshoff, T. Friemuth, C. Marzenell, Properties of honed gears during lifetime, CIRP Ann. 49 (1) (2000) 431–434.

[10] J. Masseth, M. Kolivand, Lapping and superfinishing effects on surface finish of hypoid gears and transmission errors, Gear Technology Magazine (2008) 72–78.

[11] M.P. Groover, Fundamentals of Modern Manufacturing: Materials, Processes and Systems, fourth ed., John Wiley & Sons Inc, New York, NY, 2010, p. 544.

[12] B.W. Cowley, Micro skiving: precision finishing of hardened small diameter fine module/pitch gears, splines, and serrations, Gear Solutions Magazine (2013) 65–68, September.

[13] ⟨http://www.emag.com/technologies/skiving.html⟩.

[14] G. Capello, S. Bertoglio, A new approach by electrochemical finishing of hardened cylindrical gear tooth face,, Ann. CIRP 28 (1) (1979) 103–107.

[15] C.P. Chen, J. Liu, G.C. Wei, C.B. Wan, J. Wan, Electrochemical honing of gears: a new method of gear finishing, Ann. CIRP 30 (1) (1981) 103–106.

[16] L.R. Naik, Investigation on precision finishing of gears by electrochemical honing, MTech Thesis, Mechanical and Industrial Engineering Department, Indian Institute of Technology Roorkee (India), 2008, 163 pages.

[17] J.P. Mishra, Precision finishing of helical gears by electro chemical honing (ECH) process, MTech Thesis, Mechanical and Industrial Engineering Department, Indian Institute of Technology Roorkee (India), 2009, 86 pages.

[18] G. Wei, H. Wang, C. Chen, Field controlled electrochemical honing of gears, Prec. Eng. 9 (4) (1987) 218–221.

[19] F. He, W. Zhang, K. Nezu, A precision machining of gears: slow scanning field controlled electrochemical honing, Jpn. Soc. Mech. Eng. Int. J., Ser.—C 43 (2000) 486–491.

[20] J.H. Shaikh, Experimental investigations and performance optimization of electrochemical honing process for finishing the bevel gears, PhD Thesis, Discipline of Mechanical Engineering, Indian Institute of Technology Indore (India), 2013, 134 pages.

[21] S. Pathak, Investigations on the performance characteristics of straight bevel gears by pulsed electrochemical honing (PECH) process, PhD Thesis, Discipline of Mechanical Engineering, Indian Institute of Technology Indore (India), 2016, 174 pages.

[22] V.K. Jain, Advanced Machining Processes, Allied Publishers, New Delhi, 2002.

[23] S. Santosh Kumar, S.S. Hiremath, A review on abrasive flow machining (AFM), Proc. Technol. 25 (2016) 1297–1304.

[24] L.J. Rhoades, T.A. Kohut, N.P. Nokovich, Unidirectional abrasive flow machining, US Patent Number 5,367,833, 1994.

[25] L.J. Rhoades, T.A. Kohut, Reversible unidirectional AFM. US Patent Number 5,070,652, 1991.

[26] L.J. Rhoades, Orbital and or reciprocal machining with a viscous plastic medium, International Patent No. WO 90/05044, 1990.

[27] J. Kenda, J. Duhovnik, J. Tavcar, J. Kopac, Abrasive flow machining applied to plastic gear matrix polishing, Int. J. Adv. Manuf. Technol. 71 (2014) 141–151.

[28] Y.C. Xu, K.H. Zhang, S. Lu, Z. Liu, Experimental investigations into abrasive flow machining of helical gear, Key Eng. Mater. 546 (2013) 65–69.

[29] G. Venkatesh, A.K. Sharma, N. Singh, P. Kumar, Finishing of bevel gears using abrasive flow machining, Proc. Eng. 97 (2014) 320–328.

[30] K. Gupta, N.K. Jain, R.F. Laubscher, Chapter 2 Electrochemical Hybrid Machining Processes, in Hybrid Machining Processes: Perspectives on Machining and Finishing, Springer International Publishing AG, Switzerland, 2016, pp. 9–13.

[31] V.K. Jain, Advanced Machining Processes, Allied Publisher Private Ltd, New Delhi, 2002.

[32] R. Bertsche, High pressure water deburring, Gear Solutions Magazine (2009) 20–25.

[33] B. Watkins, The electrochemical process for deburring, Gear Solutions Magazine (2005) 15.

[34] E. Mutschler, M. Nicholson, The basics of brush deburring, Gear Solutions Magazine (2004) 40–47.

[35] G. Sroka, L. Winkelmann, Superfinishing gears: the state of the art, Gear Solutions Magazine (2003) 28–33.

[36] M. Ruhland, Black oxide finishing for gears, Gear Solutions Magazine (2010) 43–47.

Chapter 6

Surface Property Enhancement of Gears

6.1 THE NEED FOR SURFACE MODIFICATION OF GEARS

Industry is continually demanding enhanced performance of gears and gear-systems to keep abreast of technological advances. Typical enhancements that are required are higher power output, lower frictional losses, reduced lubrication, lower heat, and improved reliability. In most cases, these requirements can be satisfied by surface modification of the gears.

The service life of machine parts such as gears, which are subjected to significant loads, is essentially determined by two types of tribological failures: *scuffing*, which is a severe form of mechanical wear, and *pitting*, which is a surface fatigue phenomenon [1–4].

Scuffing can be described as a localized damage caused by solid-state welding occurring between the sliding gear flanks. It is characterized by a transfer of material between the sliding contact surfaces [1,4]. Scuffing results from a local failure of the gear lubricant (i.e., the lack of continuous lubricating film) caused by frictional heating due to high sliding speed and high contact pressure. In essence, this implies that, if the lubricant film is insufficient to prevent significant metal-to-metal contact, the oxide layers that normally protect the gear-tooth surfaces may be compromised and the bare metal surfaces may weld together. This results in tearing of the welded junctions (due to "sliding" that occurs between gear-teeth), metal transfer, and catastrophic damage.

Scuffing differs from other types of gear failure such as fatigue, in the sense that it may begin at any time in a gear's life, even immediately upon start-up. Fatigue, on the other hand, develops over time depending on the load and operating conditions. Scuffing may be largely avoided by reducing the sliding velocity and minimizing the Hertzian contact stresses. Fig. 6.1A depicts an example of scuffing on the teeth of an external spur gear.

Pitting occurs due to dynamic loading that causes crack initiation and propagation. As the teeth repeatedly roll over one another, fine cracks develop at grain boundaries or inclusions, mostly beneath the surface, where the pressure is the highest. These cracks usually propagate for a short distance in a direction roughly parallel to the tooth surface. When the cracks

Advanced Gear Manufacturing and Finishing. DOI: http://dx.doi.org/10.1016/B978-0-12-804460-5.00006-7

FIGURE 6.1 Gear surface failures. (A) Scuffing on gear-teeth surfaces; (B) destructive pitting on gear-teeth surfaces.

have grown to the extent that they compromise a significant surface area, a pit is formed. Pitting is a common failure mode for gear-teeth and usually originates with micropitting that may eventually develop into severe pits. To avoid pitting, and reduce its effect on the life of a gear-set, the contact stress must be kept low, and the gear-tooth surface must have high strength (hardness) [1–3]. Fig. 6.1B shows the occurrence of severe (destructive) pitting on gear-teeth surfaces. A more in-depth discussion on the wear and failure mechanisms of gears may be found in [1].

Engineered gear-tooth surfaces with enhanced surface mechanical properties are therefore highly desirable to combat friction, reduce wear, and improve fatigue strength.

There are a number of ways to address these problems. These include:

- Diffusing carbon or nitrogen into the gear-teeth surfaces to impart necessary hardness at certain depth, thereby improving strength and fatigue properties;
- Depositing thick or thin surface layers to protect the gear-teeth during the critical running-in period;
- Introducing compressive residual stresses either by thermal or mechanical means to the gear-tooth surfaces, thereby improving the mechanical performance of the surface layer. A mechanical technique may also increase the hardness due to work hardening.

It has been demonstrated that when a loss of lubrication occurs after a period of normal lubrication, surface modifications and improvements may extend the time period of running-in before the failure occurs [3].

6.2 GEAR SURFACE MODIFICATION TECHNIQUES

The surface modification strategies discussed in the previous section can be achieved by the following specific techniques.

6.2.1 Case Hardening of Gear-Teeth Surfaces

This particular technique is used to harden the outer surface of gear-teeth in order to impart high resistance to wear, pitting, and fatigue. Case-hardening produces a hard, wear-resistant case on top of a softer, shock-resistant core. The application of gears involve dynamic forces, occasional impacts, and constant friction. The surfaces of gear-teeth therefore needs to be hard enough to prevent wear whereas the core needs to be soft enough for adequate toughness (not brittle). This can be achieved by the case-hardening process. Case-hardening of gear-teeth surfaces can be done using various heat treatment methods such as gas carburizing and vacuum carburizing type conventional processes, and advanced methods such as plasma nitriding, induction hardening, and flame hardening. The scope of this chapter limits the discussion to plasma nitriding, induction and flame hardening of gears.

6.2.2 Coating the Gear-Teeth Surfaces

This technique involves the deposition of thin layers of metal oxides, carbides, and nitrides, and even polymers and ceramics, to coat the gear-teeth surfaces. Boron carbide (B_4C), diamond-like carbon (DLC), molybdenum disulfide (MoS_2), and tungsten carbide-amorphous carbon (WC/C) coatings are most often applied for the surface modification of gears. Applications of coatings extend the running-in period, reduce the risk of scuffing and eliminate the requirements of environmentally unfriendly additives in lubricants. The deposition process can be accomplished by vapor deposition, sputtering, and thermal spraying processes. Physical vapor deposition (PVD) and chemical vapor deposition (CVD) are the two main types of vapor deposition processes. In these processes the layer (film) of deposited material is thin and usually ranges from 0.1 μm to 1 mm.

Details on various surface coatings, along with their deposition techniques, are presented in the upcoming sections.

6.2.3 Mechanical Hardening of Gear-Teeth Surfaces

This technique involves the application of mechanical impulses (e.g., light hammering) on the tooth surfaces of gears. This mechanical loading produces localized plastic flow in the near surface, and results in the work hardening of the surface layer. This also usually leads to the introduction of compressive residual stresses that is beneficial for fatigue life. This may improve the contact and bending-fatigue life of gears significantly. Some important mechanical hardening processes for gears include *shot peening* (utilizing metallic or ceramic spheres of various sizes), *shotless water jet peening* (using a jet of water with cavitation at high velocity and pressure), *ultrasonic peening* (using ultrasonically vibrated shot to impart compressive residual stresses), and *laser peening* (impacting the surface with impulses from a laser). These methods of mechanical hardening are discussed in detail in the following sections.

6.3 GEAR COATINGS

6.3.1 Introduction and Coating Types

Surface coating processes involve depositing a layer of molten, semimolten or chemical material onto a substrate. Surface coating intends to modify and reinforce the surface functions with minimal influence on the composition of the bulk material. Coatings for gears can be relatively inexpensive or prohibitively costly. Their uses range from merely cosmetic to significant improvements to the life of a gear-set. Coatings are promising antiscuffing treatments that minimize wear and enhance the service lives of gears.

There are numerous potential advantages to using surface coatings on gears [5–7]:

- Control of the running-in process;
- Prevention of scuffing or scoring with wear-resistant low-friction coatings;
- Increased rolling contact fatigue life as a result of reducing surface stress level or surface strengthening;
- Decreased noise level as a result of using a soft coating;
- Increasing gear flank load capacity;
- Excellent resistance to external environmental factors such as sunlight, salt water, technological fluids, and chemical corrosion.

6.3.2 Types of Coatings for Gears

The following are some of the available gear coatings used for low- to high-precision gearing and light to heavily-loaded gear-sets [1,5,8].

Boron carbide (B_4C) is a hard, amorphous ceramic material applied to the gear surface by magnetron sputtering PVD. Major auto makers use this coating for transmission gears. Typical coating thickness is between 2 and 3 μm.

Diamond-like carbon (DLC) is a hard, (1000–3000 HV) low-friction coating of an amorphous form of carbon with diamond-like bonds. DLC coatings basically consist of a mixture of diamond (Sp^3-tetrahedral) and graphite (Sp^2-trigonal). The relative amounts of these two phases determine many of the properties of the coating. There are numerous types of DLC coatings, including hydrogen-free (a-C), hydrogenated (a-C:H), or metal-doped (Me-C:H) coatings. DLC films doped with metal (Me-C:H) have advantages over pure carbon coatings, as internal stress is reduced and adhesion to steel substrates is improved. These coatings are applied using PVD and CVD processes. DLC coatings are also used as low-friction coatings.

They usually have high hardness and a reduced coefficient of friction (in contact with materials such as steel) and are generally chemically inert. In addition, these coatings are preferred for improving the tribological performance of gears because of their smoothness, pinhole-, and defect-free nature, and ability to provide a good diffusion barrier against moisture and gases.

Molybdenum disulfide (MoS_2) has mainly been used for dry-film lubrication for gears. It facilitates a low coefficient of friction with a high load-carrying capacity. PVD sputtering technique is used to apply this coating on high precision gears to attain tighter tolerance with thinner films. On the other hand, techniques such as spraying or dipping are used to apply these coatings on low precision gears. The performance of MoS_2 coatings can be improved with the codeposition of other metals such as titanium and chromium.

Tungsten carbide carbon coatings (WC/C) are metallic-hydrocarbon (WC-C:H) surface coatings applied by the reactive sputtering PVD process. These coatings are applied on motorcycle gears, concrete mixer gears, bevel gear actuators for aircraft landing gear flaps, and worm-gear drives. Gears that are subjected to high loads benefit greatly from these carbon-based, tungsten carbide (*WC/C*) doped coatings. These are also used as low-friction coatings.

Conversion coatings mainly include phosphate, black oxide, and electroless nickel coatings. These are used essentially as corrosion-prevention coatings. A black oxide coating is formed by a chemical reaction with the iron in a ferrous alloy to form magnetite (Fe_3O_4). It can greatly enhance the aesthetic appeal of gears. Electroless nickel coatings also offers high resistance to corrosion and wear, lubricity, and appearance benefits. Special additives such as diamond particles are used in electroless nickel coatings, for additional benefits. Phosphate coatings, more specifically manganese phosphate coatings ($MnPO_4$), are known to prevent corrosion and welding and to reduce wear. An increase in phosphate grain-size decreases friction and wear and improves the fatigue life.

Polymer coatings may provide significant improvements in lubricity and chemical resistance. Polytetrafluoroethylene (PTFE), polyvinylidene difluoride, and perfluoroalkoxy alkanes are some of the important polymers used in gear coatings. Besides offering extraordinarily low coefficients of friction and excellent corrosion protection, these coatings also have antigalling properties, abrasion resistance, low surface tension (wet-ability) and noise reduction.

6.3.3 Coating Methods

Vapor deposition-based methods, thermal spraying, and electroplating techniques are mainly utilized commercially to apply (or deposit) coatings on gear-tooth surfaces [5,6,8,9]. In the following section, only vapor deposition-based advanced coating methods are discussed.

6.3.3.1 Vapor Deposition or Thin-Film Coating Methods

Vapor deposition technology includes techniques for transforming coating materials into a vaporous state via condensation, chemical reaction, or conversion. When the vapor phase is produced by condensation from a liquid or solid source, the process is referred to as PVD. When it is produced by a chemical reaction, the process is known as CVD. These processes are typically conducted in a vacuum environment, with or without the use of a plasma. Vapor deposition processes add energy and material onto the surface only, keeping the bulk of the object relatively cool and unchanged. As a result, surface properties are modified, typically without significant changes to the underlying microstructure of the substrate.

PVD and CVD are the two main types of thin-film coating methods that are used extensively for gears. PVD techniques are limited to thin films that range from 0.1 μm to 0.1 mm, whereas CVD is used both for thin films and for coatings in excess of 1 mm [5,7].

PVD describes a variety of vacuum deposition methods used to deposit thin films by the condensation of a vaporized form of the desired film material onto gear-teeth surfaces. This occurs in the absence of any chemical reaction.

There are four main steps in any PVD reaction:

1. *Evaporation*—During this stage a high-energy source, such as a beam of electrons or ions is used to bombard a target consisting of the material to be deposited. This dislodges atoms from the surface of the target by vaporizing them.
2. *Transportation*—It consists of the movement of vaporized and dislodged atoms from the target to the substrate to be coated.
3. *Reaction*—It takes place during the transportation stage where the atoms of the target surface metal react with the appropriate gas.
4. *Deposition*—This is the process of coating built up on the substrate surface. During deposition process, some process specific undesirable reactions between target materials and reactive gases may also take place on the substrate surface.

The major variants of PVD include [5,7]:

Cathodic arc deposition, in which a high intensity electric arc is discharged at the target material that blasts ionized vapor to be deposited onto the gear-tooth surface.

Electron-beam PVD, in which the material to be deposited is heated to a high vapor pressure by electron bombardment in "high" vacuum and is transported by diffusion to be deposited by condensation on the (cooler) tooth surface.

Evaporative deposition, in which the material to be deposited is heated to a high vapor pressure by electrically resistive heating in "high" vacuum.

Pulsed laser deposition, in which a high-power laser ablates material from the target into a vapor.

Sputter deposition, in which a glow plasma discharge, usually localized around the "target" by a magnet, bombards the material, sputtering some away as a vapor for subsequent deposition.

Electron-beam evaporation and sputtering processes make coatings with fine surface topographies and therefore extensively employed for precision requirements.

CVD involves the reaction or decomposition of a precursor substance (a gas) onto the substrate. In other words the basic principle of the CVD process is that a chemical reaction between the source gases takes place in a chamber which results in the production of solid-phase material that condense the substrate surfaces [5,7]. The advantages of CVD over PVD include, relatively high deposition rates and CVD does not require as high a vacuum as PVD requires. There are three steps in any CVD reaction:

- The production of a volatile carrier compound;
- The transport of the gas, without decomposition, to the deposition site;
- The chemical reaction necessary to produce the coating on the substrate.

The process of decomposition can be assisted or accelerated via the use of heat, plasma, or other processes. Sputtering, ion plating, plasma-enhanced CVD, low-pressure CVD, laser-enhanced CVD, active reactive evaporation, ion beam, and laser evaporation, etc. are some CVD-based processes. These processes generally differ in the means by which chemical reactions are initiated and are typically classified by operating pressure. A typical CVD system consists of the following parts:

- Sources of gases and feed lines for them;
- Mass flow controllers for metering the gases into the system;
- A reaction chamber or reactor;
- A system for heating up the wafer on which the film is to be deposited; and
- Temperature sensors.

Sputtering-based PVD process is the most widely used coating method for gears and is efficiently employed for tungsten carbide carbon (WC/C), boron carbide (B_4C), molybdenum disulfide, and Ti (MoS_2/Ti) composite coatings. Another important coating method, plasma-enhanced chemical vapor deposition (PECVD), is used to deposit DLC coatings on gear-tooth surfaces. The two coating methods, magnetron sputtering-based PVD and PECVD, are extensively covered in this chapter.

6.3.3.1.1 Sputtering Physical Vapor Deposition

In this process a gas plasma discharge is set-up between two electrodes: a cathode of coating material and an anode substrate to be coated. The coating material is dislodged and ejected from the cathode surface due

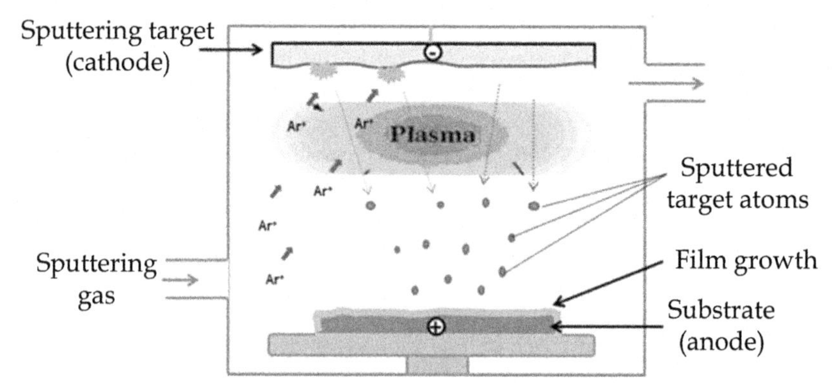

FIGURE 6.2 Sputtering process overview.

to bombardment of high-energy particles. The positive ions (and energetic neutrals) of a heavy, inert gas (e.g., argon), or a species of coating material are used as high-energy particles. The sputtered material is ejected primarily in atomic form from the cathode. The substrate (anode) is positioned in front of the target to intercept the flux of sputtered atoms (see Fig. 6.2).

The basic sputtering process has been known for many years and successfully utilized to deposit thin films on gears [1,5,6,8]. This process suffers from some inherent limitations such as low deposition rates, low ionization efficiencies in the plasma and high substrate heating effects. The development of magnetron sputtering has largely overcome the limitations of sputtering [10,11].

The use of magnetic fields to keep the plasma in front of the target and thereby intensifying the bombardment of ions makes the *magnetron sputtering* somewhat different from the general sputtering technology. A magnetic field generated due to a magnetron and mounted behind the target is used to increase the efficiency of the available electrons [10,11].

A *magnetron* is a magnetic device that is used to accelerate the process and provide greater uniformity in a PVD sputtering application. Figure 6.3 illustrates the working principle of deposition of coatings on gear by magnetron sputtering.

Magnetrons make use of the fact that a magnetic field configured parallel to the target surface can constrain secondary electron motion to the vicinity of the target. The magnets are arranged in such a way that one pole is positioned at the central axis of the target and the second pole is formed by a ring of magnets around the outer edge of the target. Trapping the electrons in this way substantially increases the probability of an ionizing electron atom collision occurring. The increased ionization efficiency of a magnetron results in a dense plasma in the target region. This, in turn, leads to increased ion bombardment of the target, giving higher sputtering rates and, therefore, higher deposition rates at the surfaces of the gear-teeth.

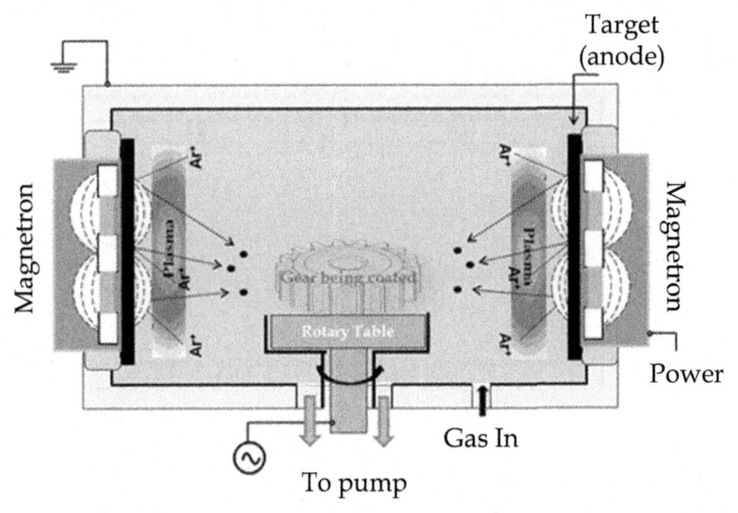

FIGURE 6.3 Schematic representation of working principle of magnetron sputtering.

In addition, the increased ionization efficiency achieved in the magnetron mode allows the discharge to be maintained at lower operating pressures (typically, 10^{-3} mbar, compared to 10^{-2} mbar) and lower operating voltages (typically, 500 V, compared to 2–3 kV) than is possible in the basic sputtering mode.

Unbalanced magnetron sputtering (UBM) is the most recent magnetron-based sputtering PVD coating technology. In this technology, some extra magnetic coils are used to intensify the plasma close to the product. This results in a denser sputtered coating without resorting to higher energies or increased temperatures.

Overall, magnetron sputtering technology is characterized by:

- High deposition rates;
- Large deposition areas;
- Low substrate heating;
- Excellent layer uniformity;
- Smooth sputtered coatings (no droplets).

This method is extensively used to coat gears with WC/C, B_4C, MoS_2, and MoS_2/Ti composite coatings, while for other DLC coatings, PECVD is the main choice.

6.3.3.1.2 Plasma-Enhanced Chemical Vapor Deposition

PECVD is used to deposit thin films of various materials on substrates at lower temperature than that of standard CVD technique.

PECVD is a hybrid coating process whereby the CVD processes are activated by energetic electrons (100–300 eV) within the plasma as opposed to thermal energy as associated with conventional CVD techniques. It is a

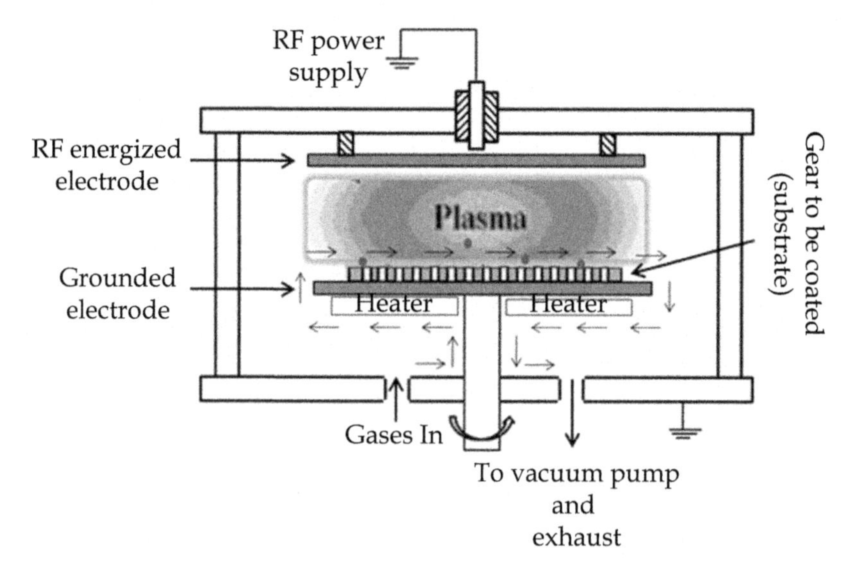

FIGURE 6.4 Gear coating inside a parallel plate radial flow reactor by PECVD.

vacuum-based deposition process operating at pressures typically <0.1 Torr allowing the deposition of films at relatively low substrate temperatures of up to 350°C.

RF-PECVD (radio-frequency discharge-based PECVD) is a plasma-enhanced CVD process where deposition is achieved by introducing reactant gases between a grounded electrode and an RF-energized electrode (see Fig. 6.4). The capacitive coupling between the electrodes excites the reactant gases into plasma, which induces a chemical reaction and results in the reaction product being deposited on the substrate, i.e., gear. The gear, which is placed on the grounded electrode, is typically heated to 300°C [12].

Since PECVD requires relatively low substrate temperatures and high deposition rates are attainable; the films can be deposited onto large-area substrates that cannot withstand the high temperatures (ranges between 600 and 800°C) such as required for traditional CVD techniques. Unlike CVD, PECVD also allows thick coatings (>10 μm) to be deposited, with a different thermal expansion coefficient, without stresses developing during the cooling period. The friction coefficient of DLC coatings, applied by PECVD technology, tends to be in the range of 0.05–0.15.

6.3.4 Testing and Inspection of Gear Coatings

The main nondestructive and destructive techniques for characterizing the properties of surface coatings include micro- and nanoindentation for hardness testing, scanned electron microscopy, and transmission electron

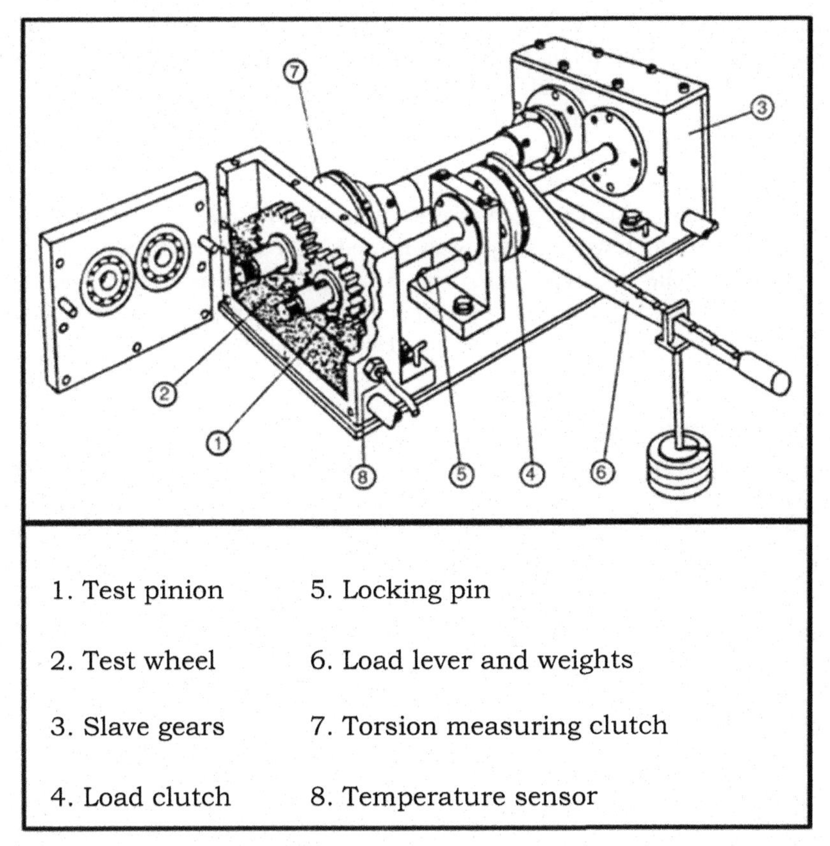

FIGURE 6.5 Schematic view of the FZG test rig [13].

microscopy for coating composition and microstructure, pin-on-disc (POD) testing for analyzing the wear rate, etc. Inspection and evaluation methods of pitted surface fatigue life and scuffing load capacity are also of considerable importance for gears. Pitting and scuff testing for gears are conducted by utilizing an apparatus referred to as the FZG gear test rig. Fig. 6.5 shows a schematic view of the FZG back-to-back gear test rig, which is used to perform gear efficiency tests and scuffing tests.

The FZG back-to-back gear test rig is a fundamental apparatus for examining the load capacity of cylindrical gears. It is mainly used to assess the fundamentals of various damage mechanisms and the associated factors of influence.

The FZG gear test rig is a back-to-back gear rig with a shaft center distance of 91.5 mm [13–16]. It accommodates two types of gears, i.e., test gears and slave gears which are connected by two shafts. One shaft is divided into two parts with the load clutch between them. One half of the load clutch can be fixed to the foundation with a locking pin, while the other half is twisted by means of a lever and weights. By removing the load and unlocking the shaft (after bolting the clutch together) a static torque is applied to the system, which can be checked in the torque measuring clutch as the twist of a calibrated torsion shaft. The rig can be operated with either a two-speed AC motor at 1500/3000 rpm, or a variable-speed DC motor between 100 and 3000 rpm.

Two standard test-gear geometries are employed for the different standard tests. For scuffing tests an A-type gear is used, with significant sliding at the pinion tip, while a C-type gear, with balanced sliding, is used for wear, micropitting, and pitting tests [14–16]. The most important test parameters, with the details of test gears, are summarized in Table 6.1.

6.3.4.1 Scuffing Test

Typically scuffing is evaluated by loading A-type gears stepwise in 12 load stages for Hertzian stresses of 150–1800 N/mm^2. Each load stage is operated for 15 min at a pitch line velocity of 8.3 m/s with an initial oil temperature of 90°C, under conditions of dip lubrication without cooling. The gear flanks are inspected after each load stage for scuffing marks by visual inspection. Failure is considered if the faces of all the pinion teeth show a summed total damage width, which is equal to or exceeds one tooth width. In the gravimetric test the gears are dismounted and weighed to determine their weight loss. Specific wear parameter can be evaluated from the graph of wear rate.

6.3.4.2 Pitting Test

During the FZG test to investigate micropitting load capacity, C-type gears, with a roughness value <1 μm are run at a pitch line velocity of $v = 8.3$ m/s. A 1-h run-in at load stage 3 (Hertzian stress 500 N/mm^2) is followed by a test in load stages 5 through 10 (Hertzian stress 800–1500 N/mm^2). Running time is 16 h per load stage. After every load stage the gears are dismounted and the profile deviation is measured. If the profile deviation exceeds 7.5 μm (corresponding to a change of DIN accuracy from 5 to 6), the test is terminated because the failure load stage has been reached. For macropitting the failure criterion is normally 4% pitted area on one tooth flank of the pinion [17].

Two more tests, namely the low-speed ZF wear test, and the ZF efficiency test to determine the gear's power loss and friction coefficient, can be conducted on a FZG test rig [9].

The major limitation of the FZG testing machine is that it only allows for a single size gear to be evaluated. This limits the versatility of the machine for testing gears of different sizes and materials.

Besides the FZG tests, some power recirculation type test rigs, with adequate arrangements of sound and vibration sensors, are utilized to perform vibration and noise tests of the coated gears.

TABLE. 6.1 Parameters and Gear Details for FZG Test [9,15,16]

Test Designation	Gear Type	Pitch Line Velocity (m/s)	Temperature (°C)	Pressure (N/mm^2)	Test Procedure
FZG scuffing test	A	8.3	90 (start temp)	150–1800	15 min/load stage (LS); and maximum 12 LS
FZG pitting test	C	8.3	90	1500	16 h/LS
ZF wear test	C	0.03	80	2035	20 + 40 + 40 h
ZF efficiency test	C	8.3	20–100	1190	About 20 min/temperature stage

Gear	Specification						
	Number of Teeth		Module (mm)	Pressure Angle (°)	Face Width (mm)	Material	Center Distance (mm)
	Pinion	Gear					
A-type	16	24	4.5	20	20	20 MnCr5	91.5
C-type	16	24	4.5	20	14	16 MnCr5	91.5

6.3.5 Past Work on Improving Tribological Characteristics of Gears Using Various Coating Types

The application of hard coatings onto gear-teeth surfaces has produced some promising results. A brief review of some notable investigations to characterize the tribological behavior of different gear coatings include the following [17–29]:

- Joachim et al. [17] studied the tribological behavior of tungsten carbon carbide (WC/C) and boron carbide coatings (B_4C) deposited by the PVD process. The thicknesses of coatings achieved were 1–4 μm for WC/C and 2 μm for B_4C, at corresponding temperatures of 150–250°C and 120°C, respectively. The FZG gear test rig was employed to evaluate the load capacity of coated and uncoated gears. A considerable improvement in scuffing load capacity was observed with coated gears. Low-speed wear tests, conducted to investigate the wear characteristics, revealed that the wear rates for coated gears were approximately five times less than for uncoated gears.
- Fujii et al. [18] studied the influence of WC/C coating on the surface durability of case-hardened steel gears. The effects of plain WC/C coating and a WC/C coating with about 1 μm CrN interlayer on the performance of a case-hardened and ground gear pair made of chromium molybdenum steel were investigated. The performance tests were carried out with a power circulation type gear test rig. Under a maximum Hertzian stress of 1700 MPa, the gears deposited with plain WC/C coating outperformed the uncoated and WC/C with CrN coated gears. Furthermore, the increasing number of load cycles represents an increase in fatigue life of gears.
- In a recent study, Tuszynski et al. [19] evaluated the performance characteristics of WC/C-coated spiral bevel gears (wheel and pinion) used in coal mines. Two testing instruments were used, namely a bevel gear laboratory test rig and an industrial gear stand equipped with chain conveyor simulating the conditions typical of coal mines. The spiral bevel gear test specimens were made of 18CrNiMo7 steel with DIN-9 quality. During the test the loading torque was gradually increased in 12 load stages (the pinion loading torque was changed from 3.3 up to 535 Nm); the power transferred by the gears varied between 1 and 168 kW, respectively. After the test was run at a particular load stage the teeth of the pinion were examined for the modes of wear, such as grooves and scuffing. The load stage was then increased up to defined failure where the damaged area reached the area of one pinion tooth. The load stage at which the previously mentioned criterion was reached is referred to as the "Failure Load Stage" (FLS) and indicates the resistance to scuffing of the tested bevel gears. The resistance to scuffing of the WC/C-coated gears (FLS = 10) was significantly improved, compared with the uncoated gears (FLS = 8). The testing with the industrial gear stand did also not result in the appearance of any surface fatigue or pitting for the meshing combination of the uncoated pinion and the

WC/C-coated wheel, whereas significant pitting could appear on the uncoated pinion and wheel combination.

- Excellent tribological performance of MoS_2/Ti and C/Cr-coated gears were observed by Martin et al. [20]. Their study reports the outcomes of several different tests conducted on coated gears. The tests include FZG scuffing tests, transfer gearbox efficiency tests and POD tests. The composite coatings were deposited using the DC magnetron sputtering technique. A significant increase in the scuffing load capacity of coated gears was observed, especially at high speeds. It was also observed that low roughness and high quality gears can promote an increase in scuffing load capacity, as comparatively improved results were achieved by applying a combination of MoS_2/Ti coating and super-finished teeth flanks (which have an average roughness of 0.3 μm) than by using coatings with normal gears. Surface coatings reduced the friction coefficient up to 49% while running at 3000 rpm. Furthermore, coated gears also improved the efficiency of the gearbox, especially at low-speed, high-torque conditions.
- Michalczewski et al. [21] conducted some performance tests, using a back-to-back gear test rig, in order to analyze the behavior of WC/C and MoS_2/Ti-coated gears. It was found that the gears coated with low-friction coatings exhibited excellent behavior under scuffing conditions. For the two coatings tested the best resistance to scuffing/scoring (failure load stage >12) was observed when both gears were coated; however, the WC/C coating gave slightly better protection against severe wear than MoS_2/Ti and only grazes, instead of scoring, were observed for WC/C. It was concluded that to increase both scuffing resistance and the fatigue life of gears, the WC/C coating should be applied on the bigger gear (wheel) and it should have a higher number of teeth than the smaller one (the pinion).
- A significant improvement in the efficiency of reduction gearbox gears was observed by He et al. after depositing MoS_2-based Ti composite coatings [22].
- An important research program conducted by Muchalczewski et al. shows that low-friction coatings such as DLC coatings can largely replace the functions of harmful antiwear, extreme pressure additives used in oils under conditions of extreme pressure. This implies an improved ecological approach to gear lubrication [23]. Case-hardened steel gears (with hardness of 60–64 HRC) were deposited with DLC coatings and evaluated using a back-to-back gear test rig. Two sets of gears were tested, namely uncoated gears lubricated with commercial oil containing additives, and coated gears lubricated with eco-oil. The same failure load stage was achieved for both combinations. Additionally, the oil temperature was lowered by 20°C and the friction was lowered by 20% for the coated gear lubricated with eco-oil when compared with the uncoated gear. Finally, it was concluded that thin, hard, low-friction coatings can ensure environment-friendly operation of gears under high loads.

- With regard to the role of coatings for improved sustainable gear operations, two additional investigations are significant: the first one reports that a reduction in noise level (measured as sound pressure level) of about 2–4 dB can be achieved by applying a bronze-graphite Br05S20Gr0.5DMO0.5 coating on gears [24]; the second investigation claims to reduce the sound pressure level by about 8 dB after depositing a 12–20 μm thick fluoropolymer coating (contains a mixture of MoS_2 and PTFE) on the helical gears [25].

These days, nanocomposite and nanostructured coatings are being considered as potential replacements for the traditional thin films for gear applications [26–29]. These coatings are free of large pits, pinholes, protrusions, and other surface and structural defects with uniform properties [27].

One example of the successful deposition of these coatings can be found in [28], where large-size gears used in polishing machines and made of stainless steel AISI 304 were coated with a CrAlN nanocomposite coating of 15 μm thickness. The coating exhibited uniformity, smoothness, and excellent bonding with the gears. Significant improvements in the service lives of CrAlN-coated gears were also found.

MoS_2-doped Ti–Al–Cr–N multilayer nanocomposite wear-resistant coatings were successfully deposited on the components of a spiroid gears using an unbalanced magnetron sputtering process [29]. Spiroid worm and wheels (made of 41MoCr11 alloyed steel) coated with Ti–Al–Cr–N elements embedded with MoS_2, showed high hardness and good friction properties.

Coating is a part of a complex tribological system; various factors such as surface roughness of the gear and base material hardness, etc. are required to be considered and optimized to reach its full potential. Coatings have the ability to reduce the need for complex transmission oil formulations and therefore also reducing the use of additives. Surface coatings can also be combined with other surface modification treatments that may enhance the advantages of both and thus achieving specific precision surface treatment requirements.

6.4 SURFACE HARDENING OF GEARS

This section describes some important mechanical hardening and case-hardening methods used to harden the gear-teeth surfaces.

6.4.1 Mechanical Hardening

6.4.1.1 Shot Peening

Shot peening is an important techniques used for mechanically hardening gear-teeth surfaces. It is a cold working surface modification process in which a stream of round hardened steel shot is propelled onto the surface of the gear-teeth to introduce work hardening along with associated compressive residual stresses in the surface layers. The aim is to improve the fatigue properties [30]. The presence of this surface compressive stress serves to retard the initiation and growth of fatigue cracks.

Shot peening is performed by accelerating shot towards the surface of the gear-tooth to be treated. In the shot peening process, compressive residual stresses are induced as a result of the localized plastic deformation that is initiated by the impacting shot on the surface. The material beneath these localized indentations resists the deformation and a compressive residual stress field is created. Typically this layer may extend to approximately 0.25 mm but is a function of the process parameters of the peening process.

The peening shot may be metallic spheres of steel (stainless steel or cast steel), or glass beads or ceramic particles. Usually their diameter ranges from 0.18 to 0.36 mm. When a group of shots/beads hits the surface, they generate multiple indentations on the surface and the tooth becomes encased in a compressively stressed layer. A tensile stress on the surface may then be replaced by this compressive layer.

The localized indentations produced by shot peening also act as small oil reservoirs which help to promote better lubrication; reduce fretting, noise, spalling, and scoring and lower operating temperature by reducing friction.

Fig. 6.6 illustrates the mechanism of shot peening in where individual shot striking the tooth's surface material acts as a tiny hammer, making a small indentation or dimple on the surface. For the dimple to be created the surface fibers of the material must be yielded in tension. Below the tooth surface the fibers try to restore the surface to its original shape, thereby producing, below the dimple, a hemisphere of cold-worked material that is highly stressed in compression. Overlapping dimples develop an even layer of metal in residual compressive stress. It is well known that cracks initiate and/or propagate with more difficulty in a compressively stressed zone. Since nearly all fatigue and stress corrosion failures originate on the surface of the gear-tooth, compressive stresses induced by shot peening result in a considerable increase in gear life.

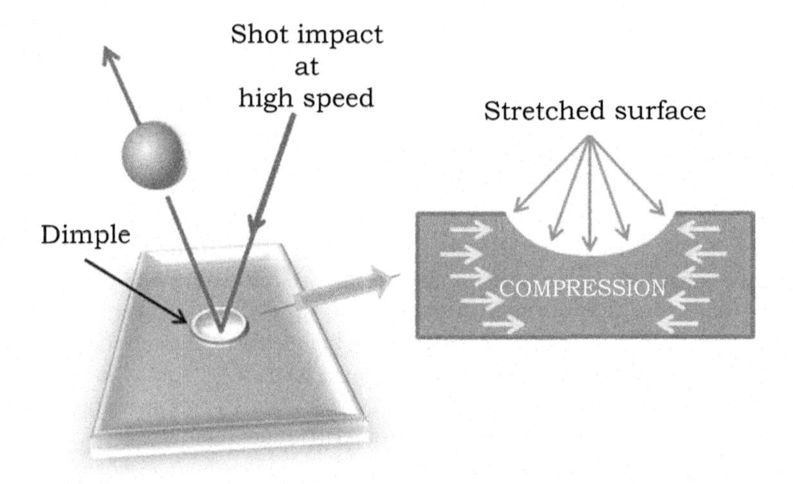

FIGURE 6.6 The mechanism of shot peening.

The shot size and type, the shot velocity, the component surface hardness, the peening exposure time, and the distance between the nozzle and the surface are the significant parameters of the shot peening process. The properties, which may be enhanced by shot peening, are fatigue strength, stress corrosion cracking, corrosion fatigue and resistance to intergranular corrosion, spalling and scoring resistance, galling and wear resistance.

Shot peening machines may be classified into two categories, namely air blast systems and centrifugal blast machines. Certain inherent limitations of the conventional shot peening have led to the development of more advanced techniques.

The major limitations of conventional shot peening techniques [30] are as follows:

- The surface finish deteriorates, resulting in high roughness of the part after ball indentation;
- Limited penetration depth and low compressive residual stresses. The penetration depth and the magnitude of the imparted compressive residual stresses are not enough to fulfill most of the application requirements;
- Long set-up time, excessive maintenance requirements and low production throughput;
- Production of large amounts of dust and waste, therefore environmentally unfriendly and implied high risk to the operator's health.

To overcome the limitations of conventional shot peening technology, and to achieve the necessary surface properties, it requires enhancements by the introduction of extended residual stress fields in a cost effective and environmentally friendly way. Conventional shot peening technology has been extensively developed and modified. Ultrasonic shot peening, laser shock peening, and water jet peening are some of the advanced types of shot peening technologies. The working principles, process mechanisms, and capabilities of these advanced shot peening methods are discussed in detail in the following sections.

6.4.1.2 Ultrasonic Shot Peening

The ultrasonic shot peening technique works on the principle of impingement of ultrasonically vibrated shot on the surfaces of gear-teeth to impart the necessary compressive residual stresses [30,31]. Because of the high frequency of the ultrasonic system the surfaces of the specimen gear-teeth to be treated are peened with a high number of impacts by only a few media (i.e., shot or balls) over a short period of time. This differs from conventional shot peening because the kinetic energy given to the media is through the acceleration of an ultrasonically vibrating surface instead of by means of air flow arrangements done to propel the balls. The frequency of the vibration is within the ultrasonic wave range, i.e., around 20 kHz.

In an ultrasonic peening system the generator generates an electrical sine wave with an ultrasound frequency. This signal is converted by a piezo-electrical emitter into a mechanical displacement, which is further amplified by booster and sonotrode. The sonotrode is attached to the part of the chamber where the gear to be treated is kept. The shots gain their kinetic energy from the sonotrode vibration and are thrown at the gear to be treated. In the case of small gears the complete gear is bombarded with high-energy shots, whereas for big gears the shots are propelled onto a specific area of the gear, covering some of its teeth. The random displacement of the balls inside the volume of the chamber ensures a uniform peening of the gear. Fig. 6.7 illustrates the working principle of ultrasonic peening of a gear.

Advances in technology have made portable solutions available in the form of a handheld, ultrasonic shot peening system/equipment for onsite shot peening treatment [31].

Advantages of the ultrasonic shot peening process over conventional peening of gears include [30,31]:

- The introduction of beneficial compressive stresses into the teeth surfaces at depths of 2–4 times that of traditional shot peening;
- Higher surface finish of the shot peened gear-teeth, therefore reduced or eliminated posttreatment polishing;
- Higher fatigue life;
- Overall shorter cycle time (20–50% cycle time reduction);
- Environmentally friendly, i.e., clean process with less dust or waste;
- No risk to the operator.

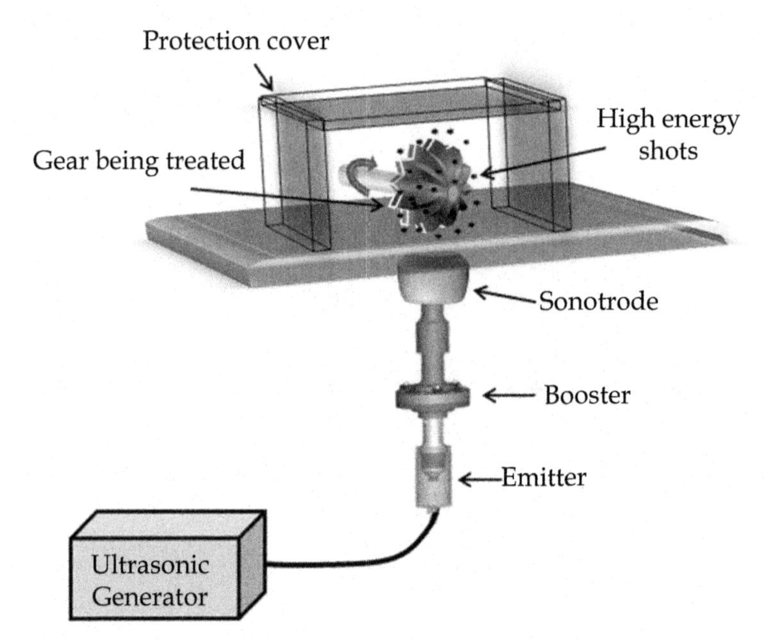

FIGURE 6.7 Schematic representation of ultrasonic peening of a gear.

6.4.1.3 Laser Peening

Laser peening is a cold working process that forms a compressive residual stress layer on gear-teeth surfaces by impacting them with a high-energy, pulsed laser beam [30]. Compared to mechanical shock peening, it is a noncontact process.

During laser peening of a gear a layer of black absorptive tape is first placed on the tooth surface of the gear to be treated in order to protect the tooth surface from the direct contact of the laser beam. A stream of water is made to flow over the surface of the tooth to be peened. The purpose of introducing the transparent layer of water is to confine the high-pressure plasma gas. An intense laser pulse is then directed at the surface to be peened.

When the laser strikes the workpiece surface, the pulse energy is absorbed by the opaque layer made by the black tape, which heats up, vaporizes and forms a high-temperature plasma [30,32]. The plasma gas is trapped between the tooth surface and the transparent water layer, limiting the thermal expansion of the gas. As a result, the gas pressure increases to an extremely high value. The high pressure is transmitted to the tooth material, producing a shock wave which travels through the tooth surface material (as shown in Fig. 6.8).

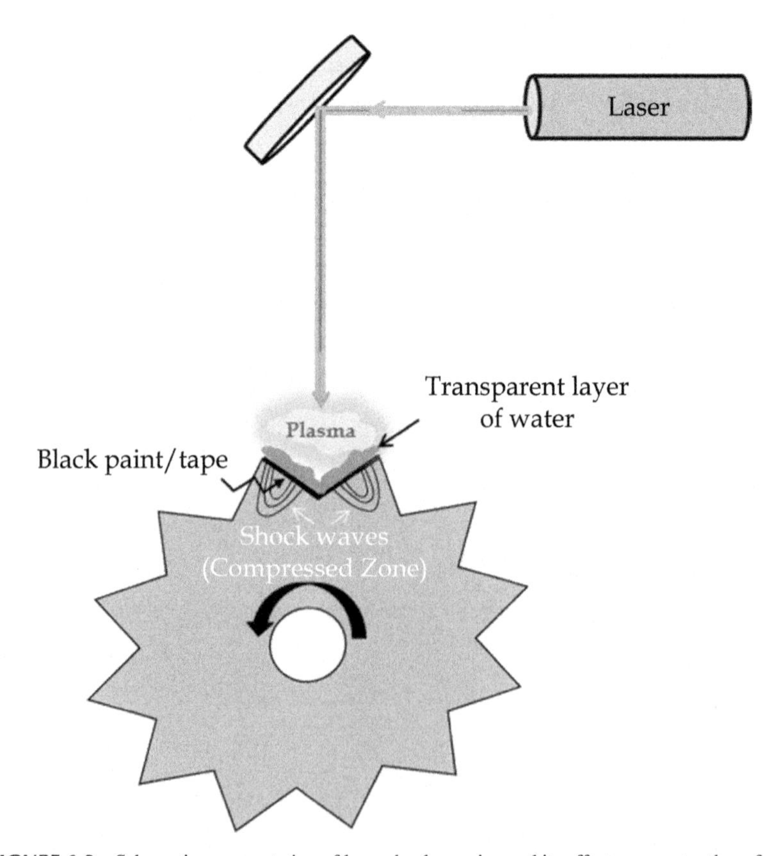

FIGURE 6.8 Schematic representation of laser shock peening and its effect on gear-tooth surface.

A compressive residual stress is then induced into the material through a subsurface shock wave.

The laser pulse can be directed onto the workpiece several times to induce an appropriate compressive residual stress field, depending on the requirements. The depth of the residual compression stress layer produced by laser peening is more than 1 mm (0.04″), which is 4–5 times deeper than the compressed zone produced by conventional shot peening.

The properties which may be enhanced by laser peening are fatigue strength, stress corrosion cracking, corrosion fatigue, resistance, and intergranular corrosion. One advantage of laser peening is that the depth of the compressive stress zone exceeds that of shot peening with less cold working (i.e., shallower indents). Owing to the deeper penetration of the compression stress zone, laser peening provides a more pronounced effect on surface related properties when compared to shot peening. The setup and running costs of laser peening are unfortunately comparatively high.

6.4.1.4 Cavitation Water Jet Shotless Peening

Cavitation water jet shotless peening is a peening technology that can produce high magnitude and deep residual compressive stresses and improves the fatigue life without much deteriorating the gear-teeth surface finish [30,33]. This process does not require shot, unlike conventional shot peening. Instead, a jet of high-velocity water with cavitation is used for peening. Fig. 6.9 illustrates the mechanism of cavitation water jet shotless peening of a gear where high-pressurized water (20–30 MPa) is injected into water through a plunger pump and a water jet peening nozzle. The high-speed water jet generates cavitation bubbles and forms a cavitation jet. Cavitation bubbles in the jet collapse on the surface of the gear-tooth, generating localized high pressure. The peening is done by the shock waves produced by the shrinking and rebounding of the cavitation bubbles, and by the impact force generated due to the microjet produced after collapsing and deforming cavitation bubbles [33,34]. The pressure exerted by the cavitation collapse is somewhere around 1000 MPa. This compressive residual stress field reduces surface residual tensile stresses in the surface and subsurface layers to mitigate stress corrosion cracking, and to improve bending-fatigue strength.

Parameters affecting the quality of peening are nozzle configuration, treatment time, flow-rate, impingement angle, and stand-off distance. Some of the gear applications require peening of the roots, some require peening of all surfaces, and some require peening at roots and teeth only. Masking is therefore done to protect the areas that are to remain un-peened.

Cavitation water jet shotless peening may offer the following benefits:

- The induced stresses are higher and deeper than with conventional shot peening;
- This process is the most cost effective of all the advanced peening techniques used for gears;

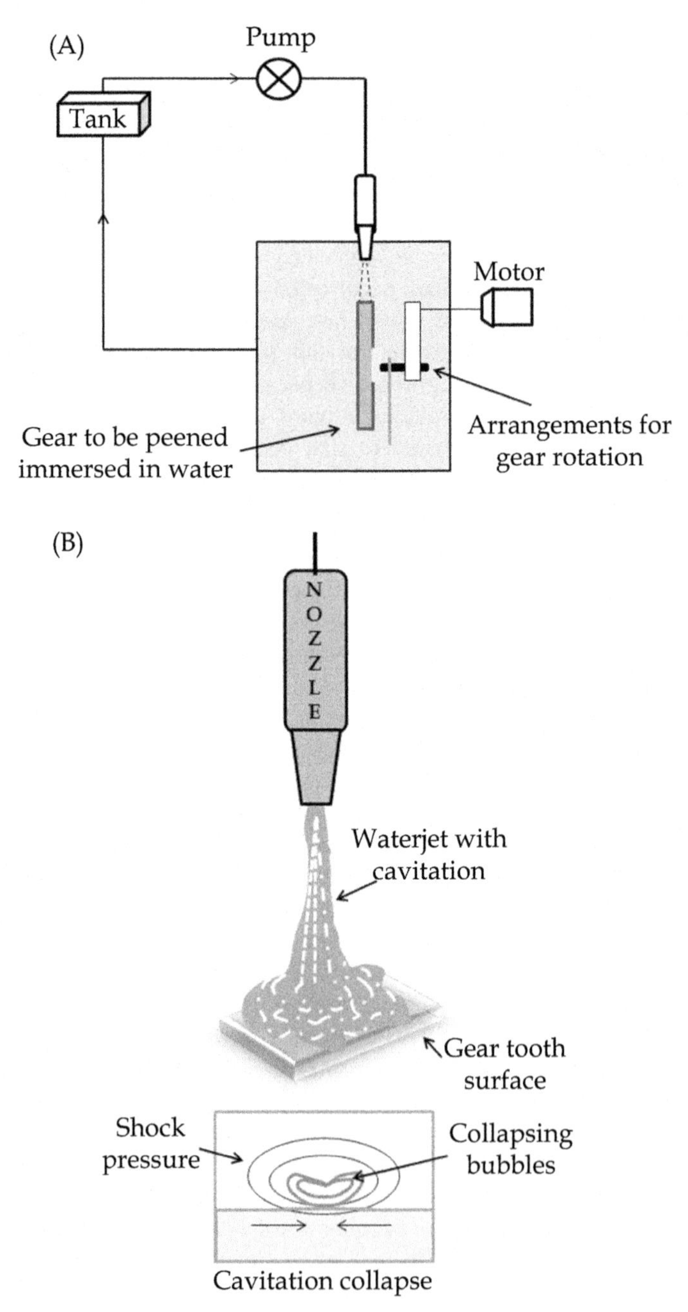

FIGURE 6.9 Cavitation waterjet shotless peening of gears (A) apparatus; and (B) mechanism.

- It is a green process, since pure water is used as a medium and therefore no waste results from the process;
- This process does not have any negative side effects such as detrimental thermal effects and there are no heat-affected zones;
- The use of a microsized nozzle can make peening of microgears possible.

Refs. [33–36] explain the complete mechanism of cavitation peening and its effect on gear-tooth surface properties. Significant efforts have been made in the past to improve the fatigue strength of gears made from materials such as carburized chromium molybdenum steel, carbon steel, aluminum alloy, magnesium alloy, stainless steel, silicon, manganese steel, and other materials [33,36]. Using water jet peening, Ju et al. successfully generated high magnitude compressive residual stresses on the helical gear-teeth made of carbon steel [36]. The uniform subsurface stress distribution was observed with maximum stress value up to 700 MPa. Soyama and Macodiyo achieved 60% improvement (relative to that of nonpeened gear) in fatigue strength of chrome–molybdenum alloy steel made spur gear after employing cavitation shotless peening [35].

6.4.2 Case-Hardening

Case-hardening is regarded as thermomechanical treatment to modify the surface properties of gear-teeth. It employs thermal diffusion to incorporate nonmetal or metal atoms into a material surface to modify its chemistry and microstructure [37]. The process is conducted in solid, liquid, or gaseous media, with one or several simultaneously active chemical elements. The mechanism of the case-hardening process includes a decomposition of solid, liquid, or gaseous species, the splitting of gaseous molecules to form nascent atoms, the absorption of atoms, their diffusion into a metallic lattice and reactions within the substrate structure to modify existing or form new phases.

Case-hardening produces a hard, wear-resistant case, or surface layer (to increase pitting resistance and bending strength) on top of a ductile and shock-resistant interior also known as core, of hardness 30–40 HRC to avoid tooth breakage [1]. Case-hardened gears provide maximum surface hardness and wear resistance and at the same time provide interior toughness to resist shock. The thickness of the hardened layer should be such that it can withstand the maximum contact stress without collapsing into the softened core of the gear-tooth. While an excessively hardened depth may make the gear-tooth too brittle to withstand tensile stresses, the exact required value of the case depth depends on the gear-tooth form and the application requirements.

The major case-hardening processes include nitriding, carburizing and their combinations, e.g., nitrocarburizing. Each of these processes has different processing requirements and technique utilizing different sources and mediums. In the last few decades, plasma nitriding, flame hardening, laser-hardening, and induction hardening methods have emerged as alternative methods for case-hardening of gears, with relatively small impact on their quality class [38–40]. These methods make use of applied energy from external sources to harden the gears. It has been claimed that these methods not only increase the wear resistance and fatigue strength of gears, but also reduce gearbox noise [40]. Some of these methods are discussed in the following section.

6.4.2.1 *Plasma Nitriding*

Plasma nitriding, also referred to as ion-nitriding, was invented by Wehnheldt and Berghause in 1932 but only became commercially viable in the 1970s. In this technique the glow discharge phenomenon is used to introduce nascent nitrogen to the surface of a gear and its subsequent diffusion into subsurface layers [37]. The process of the plasma-nitriding of gears is carried out in a furnace where an electrical voltage is applied between the gear to be treated, as the cathode, and the furnace as the anode (see Fig. 6.10).

First, the vessel is evacuated. Gas containing a mixture of nitrogen and hydrogen, occasionally enriched with argon or methane, is then introduced. Pressure in the range of 0.1–10 Torr is set to create a vacuum. At this point the electrical discharge is switched on and a glow discharge takes place; the nitrogen ion thus produced strikes the surface of the cathode with high kinetic energy, emitting heat that results in a sputtering of the cathode, which atomizes the cathode (gear) surface material. Then iron nitride is formed when the atomized ions combine with nitrogen ions in the plasma, which is

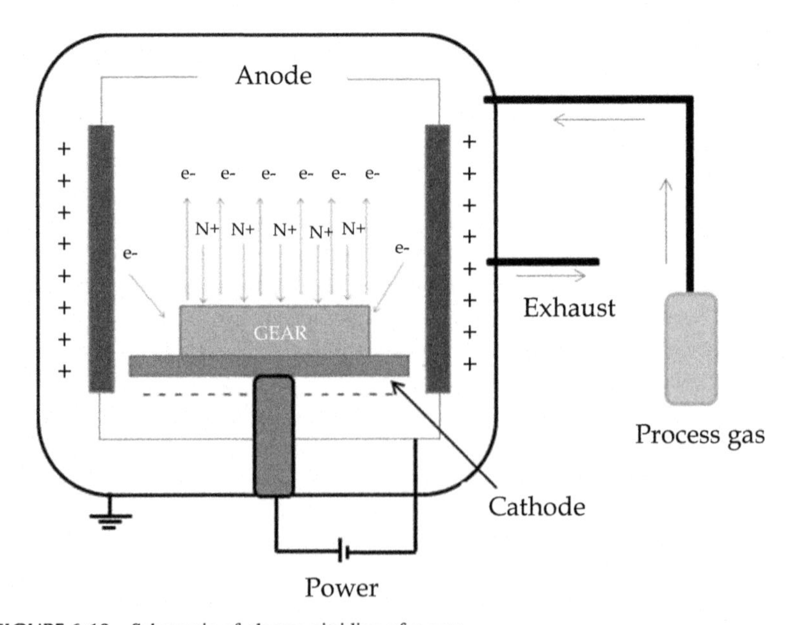

FIGURE 6.10 Schematic of plasma nitriding of a gear.

then deposited in an even iron nitride layer on the cathode. In turn the iron nitrides are partially broken down on the surface of the cathode, after which the nitrogen diffuses into the gear material and results in nitriding [41].

Plasma nitrided gears have case hardness of between 58 and 63 HRC and possess excellent wear resistance and extended service life. In addition, the fatigue strength of a gear-tooth may also be significantly increased. The formation of the precipitates on the gear-case results in lattice expansion. In order to maintain its original dimensions the core keeps the nitrided case in compression. This compressive stress lowers the applied tensile stress on the gear material, increasing the fatigue strength.

The nitrided layer has a diffusion zone and a compound zone, i.e., a white layer. In the diffusion zone, nitrogen diffuses in steel, producing a hardened zone by precipitation and solid-solution hardening. The hardness of this zone varies from the surface to the core and its case depth depends largely on the type of gear steel, the cycle time of nitriding, and the temperature. The case depth may be as high as 900 μm. The excess nitrides diffuse into the gear material during the heat-treating cycle, leaving a white layer on the top surface. This white layer is brittle and relatively inert. Its thickness is usually below 13 μm [41], which can be reduced further by controlling the ratio of nitrogen in the mixture of nitrogen and hydrogen during ion-nitriding. The composition of the white layer provides natural lubricity and corrosion resistance, provided its depth does not exceed 10–12 μm [41,42].

Benefits of plasma nitriding over traditional carburizing and nitriding include:

1. Increased gear-case hardness;
2. Improved control of case thickness and greater uniformity;
3. No distortion, therefore no postprocessing is required;
4. Can be performed at relatively low temperatures (450–550°C);
5. Environmentally friendly;
6. Improved process time compared with tradition nitriding.

6.4.2.2 Induction and Flame Hardening

6.4.2.2.1 Induction Hardening

Induction hardening is a localized heat treatment used to improve the fatigue life, strength, and the wear resistance of gears. In this method, instead of heating the whole gear; the heat can be precisely localized to the specific areas where metallurgical changes are desired (e.g., flank, root, and gear tip can be selectively hardened) and therefore the heating effect on adjacent areas is minimal. The maximum attainable surface hardness with induction hardening is about 55 HRC [39,41].

In induction surface hardening the heat input into the gear is achieved with the introduction of eddy currents. These eddy currents are the result of rapid magnetic field changes introduced by alternating electric currents in specialized conducting coils. The gear is placed inside a coil, and when a high-frequency alternating current is passed through the coil, rapid heating takes place due to electromagnetic induction. An immersion quench tank or spraying water through jets passing through the inductor coils is used to quench gear.

Selective heating and, therefore, hardening, is accomplished by designing suitable coils or inductor blocks. The depth to which the heated zone extends controls the surface hardness and case depth, and depends on the frequency of the current and on the duration of the heating cycle.

The required hardness profile and resulting gear strength and residual stress distribution are basically determined by the type of gear material (steel or cast iron), its prior microstructure and the desired gear performance characteristics. Gears made of cast irons (ductile, malleable, and gray), and low-alloy and medium-carbon steels with 0.4–0.55% carbon content (i.e., AISI 4140, 4340, 1045, 4150, 1552, 5150) are commonly heat treated by induction hardening process. Alloy steels with more than 0.5% carbon are susceptible to cracking [1,41]. External spur and helical gears, worm gears, bevel and internal gears, racks, and sprockets are typically induction hardened.

Induction hardening of gears is done by two methods: spin hardening and tooth-to-tooth or contour hardening [1]. For the case-hardening of gears subjected to high loads, contour hardening is used (see Fig. 6.11). This method is generally not applicable for a tooth size finer than 16 DP. Contour type induction hardening is performed with a shaped intensifier that oscillates back and forth in the gear-tooth space. The gear is then rapidly submerged and quenched. The process is accomplished by hardening one tooth root at a time. After each root has been hardened, the system indexes the gear to the next position and the process begins again. This process is time-consuming; however, it is used to harden large gears because their heat treatment by conventional processes requires a large amount of power and very large diameter gears are not fit inside any existing carburizing furnaces. Contour hardening provides strength and wear resistance on the contact areas of the gear while minimizing dimensional movement by leaving the tooth tips unhardened [43].

To achieve greater depth of heat penetration, low-frequency current is used, whereas heat treatment at shallow depth requires high-frequency current [1,41].

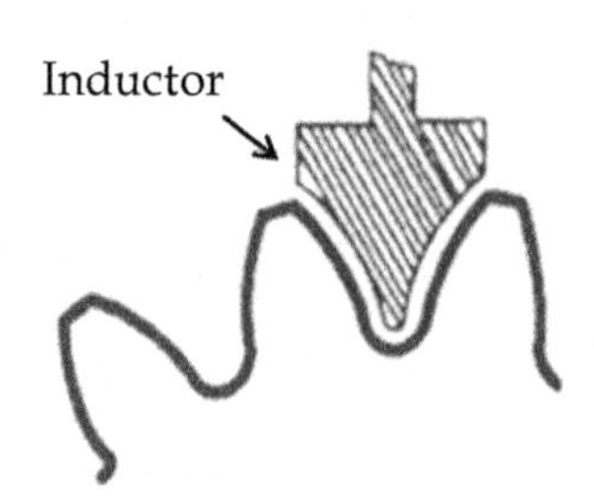

FIGURE 6.11 Tooth-to-tooth induction hardening.

The *dual-frequency process* is a recent version of induction hardening, where two different frequencies, high and low, are simultaneously used for heating. At first the gear is heated with a relatively low-frequency source (3–10 kHz), providing the energy required to preheating the extended mass of the gear-teeth. This step is followed immediately by heating with a high-frequency (100–270 kHz) source, which rapidly heats the entire tooth contour surface to an appropriate hardening temperature. Thereafter quenching of gear is done to the desired hardness [44].

The amount of heat applied by the dual-frequency process is considerably less than a single-frequency process. Therefore induction hardening by the dual-frequency method is more favorable towards generating compressive residual stresses compared to the single-frequency induction method. Furthermore, the heat treatment distortion is significantly lower in dual-frequency method. This method is particularly useful for higher root hardness and close control of case depth.

6.4.2.2.2 Flame Hardening

Flame hardening is a heat treatment process where oxyfuel gas flames are directly impinged onto the gear-tooth surface area to be hardened which is then subjected to quenching. It results in a hard surface layer of martensite over a softer interior core. Its cost is considerably less than induction hardening. Very large gears whose quality requirement is generally below AGMA class 7, and heat treatments by conventional processes are not practical or economical are flame hardened [41,45].

Basically two types of flame hardening techniques are in use, namely, spin hardening and tooth-at-a-time methods. Spin hardening is best suited for gears with enough mass to absorb the excessive heat applied in this method without too much distortion. In the tooth-to-tooth method the gear is heated and quenched by the machine itself, which limits the amount of heat going into the gear. There are two ways of heating gear-teeth. One is the tooth-to-tooth method shown in Fig. 6.12A, where the flame head provides both flank and root hardening. The other method is shown in Fig. 6.12B where only the flank is hardened, leaving the root area untreated [45].

Gases used for flame heating are acetylene and propane. Each of these gases is mixed with air in particular ratios and burnt under pressure to generate the flame that the burner directs onto the workpiece. The usual range of carbon content required in gear steels for flame hardening is 0.40–0.50%. Quenching is done either by water spray or by air. A typical hardness range for various compositions of gear steels obtained by water quenching is 45–65 HRC, and by air quenching is 45–63 HRC [1].

Modernization in surface engineering has advanced the gear industry with improved functional performance and an enhanced service life of gears. This has been done by developing advanced coatings such as hybrid- and

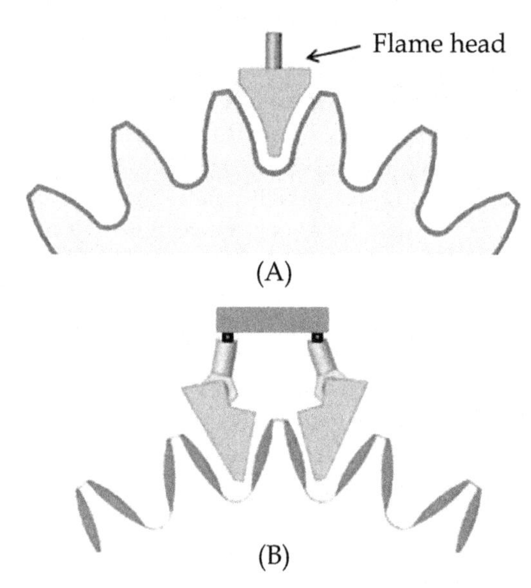

FIGURE 6.12 Tooth-at-a-time flame hardening. (A) Root/flame method. (B) Flank method.

nanocoatings and their deposition methods, which include magnetron sputtering and plasma-enhanced deposition; modern mechanical hardening methods such as ultrasonic, laser and cavitation jet peening; and advanced case-hardening methods like plasma nitriding and induction and flame hardening. These advanced surface property enhancement methods are significant improvements on conventional methods as they are fast, accurate, and efficient and do not affect the gears adversely. These advanced methods are more environmentally friendly, safer, and cleaner. They not only improve hardness, wear resistance, and fatigue strength, but also control friction, reduce adhesion, and improve lubrication and corrosion resistance. This, consequently, improves the tribological performance, load-carrying ability, and smooth motion transmission characteristics of gears, with an appreciable reduction in noise, vibration, and wear rate. Furthermore, the combination of two techniques, such as hardening and coating, or superfinishing and coating, or finishing and hardening, achieves a significantly enhanced effect than a single treatment.

REFERENCES

[1] J.R. Davis, Gear Materials, Properties, and Manufacture, first ed., ASM International, OH, 2005.

[2] P.J.L. Fernandes, C. Mcduling, Surface contact fatigue failures in gears, Eng. Failure Anal. 4 (2) (1997) 99–107.

[3] M. Murakawa, T. Komori, S. Takeuchi, K. Miyoshi, Performance of a rotating gear pair coated with an amorphous carbon film under a loss-of-lubrication condition, Surf. Coat. Technol. (1999) 120–121.

[4] T. Burakowski, M. Szczerek, W. Tuszynski, Scuffing and seizure—characterization and investigation, in: G.E. Totten, H. Liang, (Eds.), Mechanical Tribology. Materials, Characterization, and Applications, Marcel Dekker, Inc, New York, 2004, pp. 185–234.

[5] K. Holmberg, A. Matthews. Coatings Tribology: Properties, Mechanisms, Techniques and Applications in Surface Engineering, Tribology Series 28, Elsevier, Amsterdam, 1994, 442 pp.

[6] D.T. Quinto, PVD coatings for improved gear production, Gear Solutions Magazine, January 2004, pp. 24–27.

[7] S. Frainger, J. Blunt, Engineering Coatings-Design and Application, second ed., Abington Publishing, Abington, UK, 1998.

[8] W.R. Stott, Myths and Miracles of gear coatings, Gear Technology Magazine, July–August 1999, pp. 36–44.

[9] F. Joachim, Influence of coatings and surface improvements on the lifetime of gears, in: Proceedings of COST 532 Conference, Triboscience and Tribotechnology, Gent, Belgium, 2004, pp. 138–147.

[10] S.M. Rossnagel, Sputter deposition, in: W.D. Sproul, K.O. Legg (Eds.), Opportunities for Innovation: Advanced Surface Engineering, Technomic Publishing Co, Switzerland, 1995.

[11] P.J. Kelly, R.D. Arnell, Magnetron sputtering: a review of recent developments and applications, Vacuum 56 (2000) 159–172.

[12] B. Bhushan, Springer Handbook of Nanotechnology, third ed., Springer Verlag, Heidelberg, 2010.

[13] R.C. Martins, P.S. Moura, J.O. Seabr, MoS2/Ti low-friction coating for gears, Tribol. Int. 39 (2006) 1686–1697.

[14] ASTM D5182-97, Standard Test Method for Evaluating the Scuffing (Scoring) Load Capacity of Oils, 2014.

[15] Gears-FZG test procedures-Part 1, FZG test method A/8,3/90 for relative scuffing load-carrying capacity of oils (ISO 14635-1), 2000.

[16] FVA Information Sheet No. 54/I-IV, Test procedure for the investigation of the micro-pitting capacity of gear lubricants, July 1993.

[17] F. Joachim, N. Kurz, B. Glatthaar, Influence of coatings and surface improvements on the lifetime of gears, Gear Tehnology Magazine, July/August 2004, pp. 50–56.

[18] M. Fujii, M. Seki, A. Yoshida, Surface durability of WC/C-coated case-hardened steel gear, J. Mech. Sci. Technol. 24 (2010) 103–106.

[19] W. Tuszynski, M. Kalbarczyk, M. Michalak, R. Michalczewski, A. Wieczorek, The effect of WC/C coating on the wear of bevel gears used in coal mines, Mater. Sci. (Medžiagotyra) 21 (3) (2015) 358–363.

[20] R. Martins, R. Amaro, J. Seabra, Influence of low friction coatings on the scuffing load capacity and efficiency of gears, Tribol. Int. 41 (2008) 234–243.

[21] R. Michalczewski, M. Szczerek, W. Tuszyński, J. Wulczyński, M. Antonov, The effect of low-friction PVD coatings on scuffing and pitting resistance of spur gears, Tribologia 21 (5) (2013) 55–66.

[22] H. He, S. Lyu, C. Her, Effect of MOS_2-based composite coatings on tribological behavior and efficiency of gear, in: Proceedings of ASME 2010 10th Biennial Conference on Engineering Systems Design and Analysis, Istanbul, Turkey, July 12–14, 2010, vol. 1, Paper No. ESDA2010-24464, pp. 425–430.

[23] R. Michalczewski, W. Piekoszewski, M. Szczerek, W. Tuszynski, Scuffing resistance of DLC-coated gears lubricated with ecological oil, Estonian J. Eng. 15 (4) (2009) 367–373.

[24] U.L. Basiniuk, M.A. Levantsevich, N.N. Maksimchenko, A.I. Mardasevich, Improvement of triboengineering properties and noise reduction of tooth gears by cladding functional coatings on working surfaces of interfaced teeth, J. Friction Wear 34 (2013) 575–582.

[25] Z.I. Kork, G.R. Gillich, I.C. Mituletu, M. Tufoi, Gearboxes noise reduction by applying a fluoropolymer coating procedure, Environ. Eng. Manage. J. 14 (6) (2015) 1433–1439.

[26] M. Roy, Nanocomposite films for wear resistance applications, in: M. Roy (Ed.), Surface Engineering for Enhanced Performance Against Wear, Springer-Verlag, Wien, 2013.

[27] P.M. Martin, Handbook of Deposition Technologies for Films and Coatings, third ed., Elsevier, Oxford, 2010.

[28] L. Wang, G. Zhang, R.J.K. Wood, S.C. Wang, Q. Xue, Fabrication of CrAlN nanocomposite films with high hardness and excellent anti-wear performance for gear application, Surf. Coat. Technol. 204 (21–22) (2010) 3517–3524.

[29] G. Strnad, D. Biro, V. Bolos, I. Vida-Simiti, Researches on nanocomposite self-lubricated coatings, Metalur. Int. 14 (2009) 121–124.

[30] M.K. Kulekci, U. Esme, Critical analysis of processes and apparatus for industrial surface peening technologies, Int. J. Adv. Manuf. Technol. 74 (2014) 1551–1565.

[31] http://www.sonats-et.com/page_25-ultrasonic-shot-peening.html.

[32] L. Hackel, Shaping the future-laser peening technology has come of age, Shot Peener 19 (3) (2005) 3.

[33] M. Seki, H. Soyama, M. Fujii, A. Yoshida, Rolling contact fatigue life of cavitation-peened steel gears, Tribol. Online 3 (2) (2008) 116–121.

[34] H. Soyama, Cavitation S Peening, The Shot Peener, Summer 2014, pp. 16–20.

[35] H. Soyama, D.O. Macodiyo, Fatigue strength improvement of gears using cavitation shotless peening, Tribol. Lett. 18 (20) (2005) 181–184.

[36] D.Y. Ju, H. Tsuda, V. Ji, T. Uchiyama, R. Oba, Residual stress improved by water jet peening for a quenched gear, in: Transactions of Materials and Heat Treatment Proceedings of the 14th IFHTSE Congress, October 2004, 25 (5) 502–507.

[37] F. Czerwinski, Thermochemical treatment of metals, in: F. Czerwinski (Ed.), Heat Treatment-Conventional and Novel Applications, InTech, Croatia, 2012, pp. 73–111.

[38] D.H. Herring, G.D. Lindell, Heat treating heavy duty gears, Gear Solutions Magazine, October 2007, pp. 59–75.

[39] F.J. Otto, D. Herring, Heat treat 101: a primer, Thermal Processing for Gear Solutions, April 2014, pp. 40–44.

[40] M.A.K. Babi, Gears-the art of heat treating high precision gears, Modern Manufacturing India Magazine, March 2013, pp. 42–44.

[41] A.K. Rakhit, Heat Treatment of Gears, first ed., ASM International, OH, USA, 2000.

[42] http://www.parkermotion.com/engineeringcorner/gearheads.html.

[43] N. Bugliarello, B. George, D. Giessel, D. McCurdy, R. Perkins, S. Richardson, et al., Heat treat processes for gears, Gear Solutions, July 2010, pp. 40–51.

[44] J.M. Storm, M.R. Chaplin, Dual frequency induction gear hardening, Gear Technology, March/April 1993, pp. 22–25.

[45] M.M. Sirrine, Flame hardening, Gear Solutions Magazine, October 2015, pp. 69–78.

Chapter 7

Measurement of Gear Accuracy

Gear metrology refers to a special branch of metrology as relevant to gear accuracy, gear measurement techniques, and instruments. It usually requires specialized training and expertize to perform adequately.

The main objectives of gear accuracy measurement are as follows:

- Check the compatibility/level of gear accuracy with the specified tolerances before actual use, i.e., to assure required accuracy and quality.
- Provide an insight into the performance of the gear manufacturing process including the setup of the gear making machine tools, condition of the gear cutting tools, machine tool control, and basic machining practices.
- Determine the distortions caused by possible heat treatment to facilitate corrective action.
- Minimize overall cost of manufacture by controlling rejection and scrapping.

In other words, gear inspection and measurement reveals the level/amount of deviations in geometry and hence the quality class of gears, conditions of gear manufacturing machines, current state of gear cutting tools, possible sources and causes of errors, and informs about the requirements of process and/or machine tool control and other pre- and posttreatments to be conducted to fulfill the accuracy requirement pertaining to the specific application.

Various parameters defining gear accuracy, gear quality standards, aspects of gear metrology, and some common accuracy measurement/inspection methods are discussed in the following sections.

7.1 GEAR ACCURACY

Gear accuracy refers to how closely its main geometric features resemble the theoretical design. Gears are complex geometric shapes and are therefore specified by a range of appropriate dimensions. Gear accuracy may then be considered as the deviation of an intelligent combination of selected dimensions from the theoretical design. These dimensions are then also evaluated within certain tolerance levels. To simplify this process, it is therefore important to qualify the range of these tolerance levels in a more general sense by way of some standard or code of practice usually by way of a systematic

Advanced Gear Manufacturing and Finishing. DOI: http://dx.doi.org/10.1016/B978-0-12-804460-5.00007-9

numbering scheme. The gear accuracy or quality may then be described by a single number that indicates how closely it complies with an appropriate acceptance standard.

For applications that require high accuracy, it is usually mandatory to inspect and qualify the gear before actual use. Operation requirements of gears may include, specified maximum torque transfer capability, minimum running noise, and accurate rotation/positioning that are only possible with an appropriate degree of gear accuracy. Close monitoring of gear accuracy during its manufacture is therefore not only important for operational requirements but also for detecting and monitoring problems during manufacturing as related to the machine tools and their operation.

Significant gear metrology parameters that are used to quantify gear accuracy may be divided into two major classes [1]:

1. Dimensional or macrogeometry parameters; and
2. Microgeometry parameters.

Inaccuracy or errors in the micro- and macrogeometry of a gear causes deviation from the ideal motion transmission conditions. In other words, the level and amount of deviations/errors in these parameters govern the functional performance of gears.

7.1.1 Macrogeometry Parameters

It could be argued that the most significant macrogeometry parameter of a gear is the tooth thickness. The tooth thickness is generally defined as the length of arc of the pitch circle between opposite faces of the same tooth, i.e., the thickness of a gear tooth along the pitch circle [1,2]. The error or deviation in tooth thickness is the difference between its theoretical and actual values. It largely governs the functional characteristics of the gears which in turn affect their operating performance. Most notably, tooth thickness errors are the main cause for excess or reduced backlash between the mating gears. Excess backlash may cause noise (on reversal) but also be responsible for a reduction in effective tooth strength. Tooth thickness can be monitored by physical measurement using a gear tooth Vernier calliper; a micrometer used along with balls, pins, or rolls; and plate micrometers. It can also be measured by functional inspection.

Other macrogeometry parameters closely related to the tooth thickness are chordal tooth thickness, span, and diameter over balls [3].

Chordal tooth thickness is the tangential gear tooth thickness and defined as $0.5(\pi \times \text{module})$. It is easier to measure directly than the gear tooth thickness. The deviation in chordal tooth thickness is the difference between its theoretical and actual values.

Span is the distance across a certain number of teeth along a line tangent to the base circle. The length of span is composed of base pitches and tooth thickness, and hence the value of tooth thickness can be calculated using a formula interrelating them.

Diameter over two balls is the maximum distance between diametrically opposed tooth spaces of a gear and measured when two balls, pins, or rolls of the same size are placed inside these tooth spaces.

7.1.2 Microgeometry Parameters

The two major classes of microgeometry parameters are *form parameters* and *location parameters.* The former involves the shape of the teeth while the latter is associated with the actual teeth positioning [1].

Significant form parameters are profile and lead, whereas pitch and run-out are significant as far as location parameters are concerned. The number and magnitude of errors in these microgeometry parameters quantifies the quality of gears and may significantly affect their functional performance. A schematic of the various microgeometric errors and their effects are presented in Fig. 7.1. Deformation and inaccurate clamping of the workpiece and tool, form defects of the cutter, vibration in the machine tool, errors in the machine tool axis, etc. are notable causes of errors in microgeometry parameters. In general, the assignment of gear quality grades are a function of the severity of the form and location errors.

7.1.2.1 Profile and Profile Errors

The profile is that section of a cross-sectional view of the tooth that lies between the tip or outside diameter of the tooth and the root (see Fig. 7.2). The portion of the profile that is actually in contact with its meshed partner gear is of special significance and the most important focus of gear metrology. It is referred to as the functional or active profile. The function profile lies between the profile control diameter (just above the base circle diameter) and the addendum circle diameter or the start of tip correction (see Fig. 7.3). The corresponding points are known as "start active profile (SAP)" and "end active profile (EAP)," respectively.

The difference between the measured and the specified functional profile is referred to as the profile error. In other words, it refers to the deviation of the actual involute from the theoretical involute, as shown in Fig. 7.4. It has two components namely the profile form error and profile angle/slope error which are measured perpendicular to the functional profile.

Profile form error (ff_α) is the difference between the nominal involute form and the actual involute form, while *profile angle error* (fh_α) is the difference between the nominal involute angle and the actual involute angle. The *total profile error* (F_α) is the sum of profile form and profile angle errors (see Fig. 7.1).

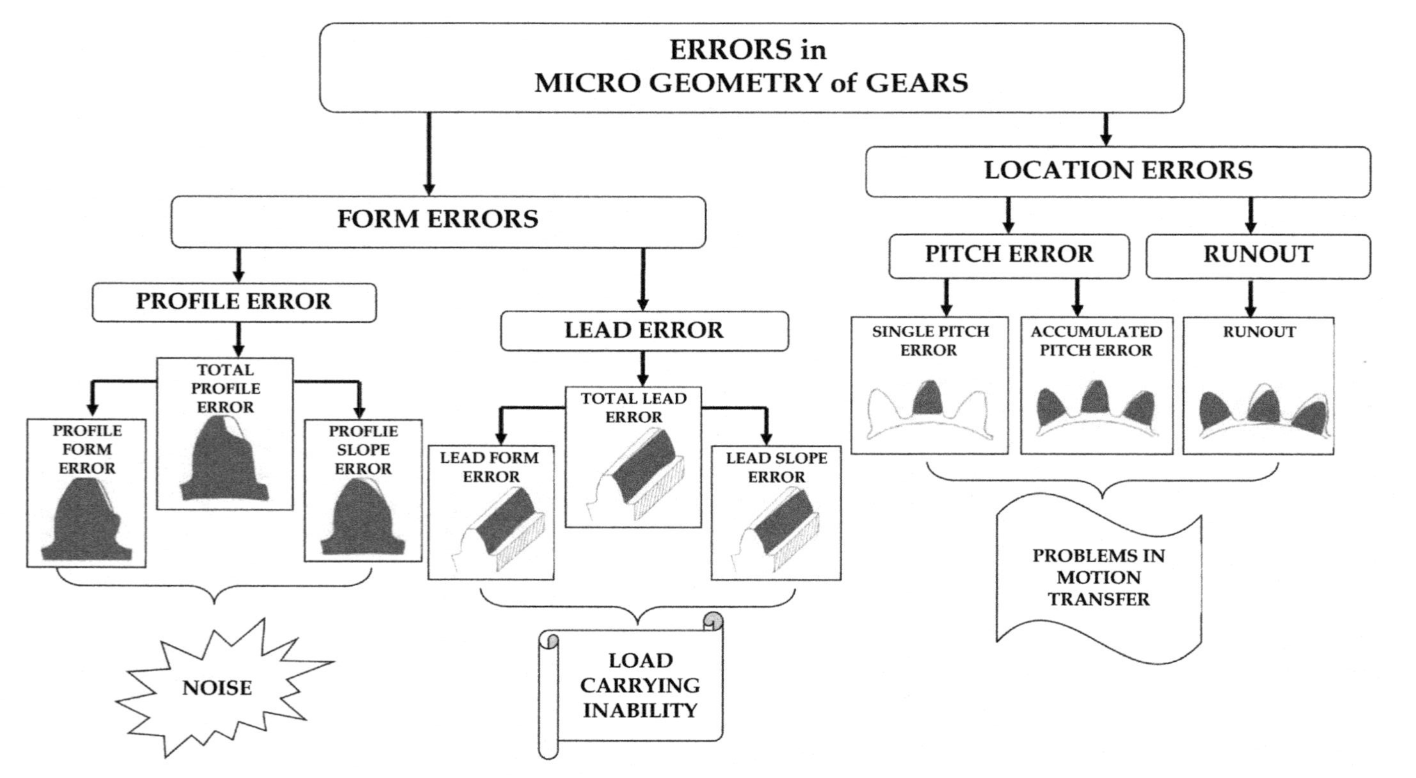

FIGURE 7.1 Schematic of microgeometry errors and their influence on gear performance [1].

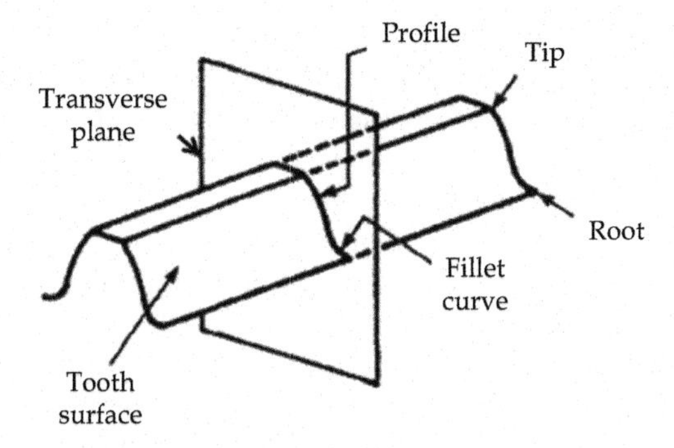

FIGURE 7.2 Representation of a gear tooth profile.

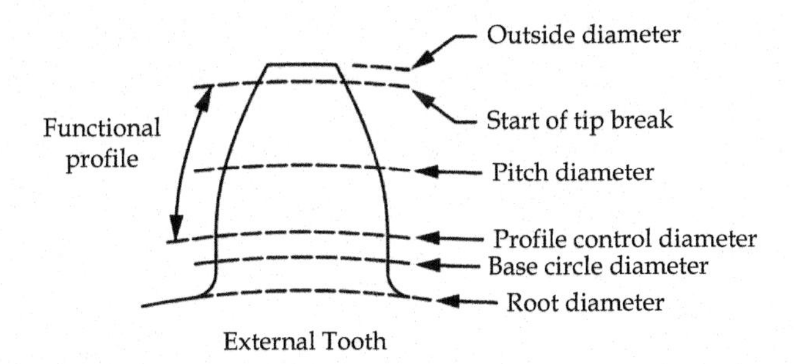

FIGURE 7.3 Representation of the functional profile.

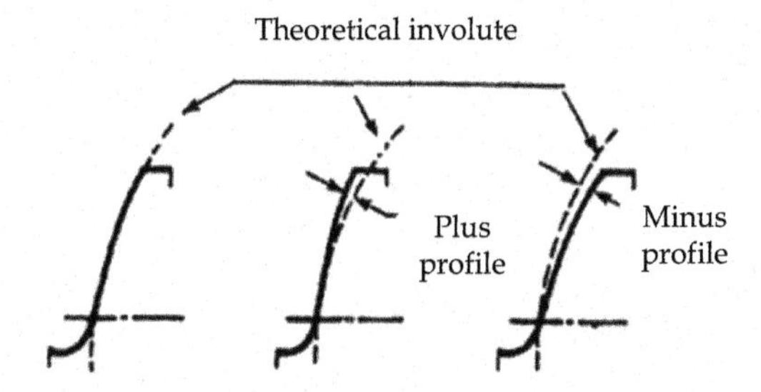

FIGURE 7.4 Concept of profile error [4].

The extent of the profile error is largely responsible for the noise characteristics of a gear [2]. Involute profile errors can also adversely affect the strength and durability of a gear by inducing localized contact stresses that may adversely affect wear and fatigue properties [4]. Geometric inaccuracies in the gear blank and cutting tool, and mounting errors are the primary causes of profile errors.

7.1.2.2 Lead and Lead Errors

Lead is the axial advance of a helix for one complete rotation of the gear. Generally, this term is used for helical gears. The lead of a helical gear is commonly specified by the angle of inclination of the helix to the axis of rotation at a specified diameter (see Fig. 7.5). Similarly, the lead may also refer to the lengthwise alignment of the tooth flank along the face width from one end to the other. The theoretical lead of a spur gear is a straight line parallel to its rotating axis.

Lead error is the difference between the specified and the measured tooth alignment of the gear [1]. It is measured in a direction normal to the specified alignment (see Fig. 7.6). The two most significant components of lead error are *lead form error* (ff_β) and *lead angle/slope error* (fh_β) that are measured at the middle of the tooth height along the face width of the gear. *Lead form error* is the difference between the nominal lead form line and the actual form line, whereas *lead angle error* is the difference between the nominal helix angle and the actual helix angle. The *total lead error* (F_β) is the sum of lead form and lead angle errors (see Fig. 7.1).

Lead measurement is used to determine correct face contact between mating gears. Errors in lead create uneven loading, nonuniform motion transmission,

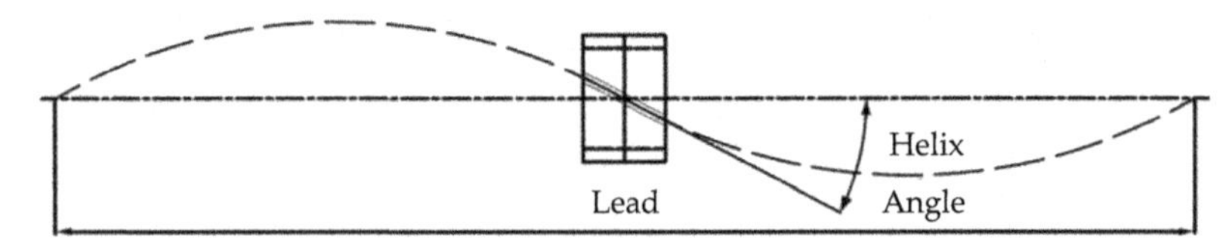

FIGURE 7.5 Lead of a helical gear.

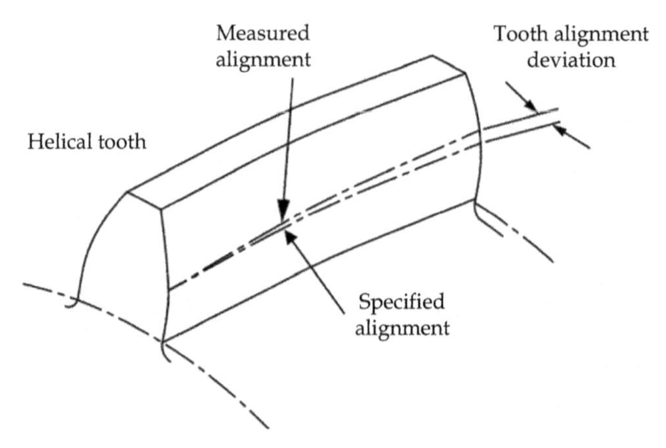

FIGURE 7.6 Tooth alignment error.

and localized loadbearing leading to the accelerated wear. The extent of the lead error is mainly concerned with the torque transfer capability of a gear [2].

7.1.2.3 Pitch Error and Runout

Pitch error and runout are the location errors which define the accuracy of location or position of teeth on a gear and hence determine the transmission accuracy [1,2]. Pitch is the distance between two corresponding points on two consecutive gear flanks on the same side along the pitch circle. The positive or negative difference between the theoretical pitch and the measured pitch in the transverse plane is the pitch error. *Pitch error* refers to the inaccuracy in angular location of the gear teeth along the pitch circle of the gear. Pitch error is mainly of two types, i.e., single pitch error (f_p) and accumulated or total pitch error (F_p) and are measured at the middle of the tooth height along the pitch circle. *Single pitch error* is the algebraic difference between the theoretical and actually measured values of the pitch for a pair of teeth as depicted in Fig. 7.7. *Accumulated or total pitch error* is the maximum value of location inaccuracy between any two teeth of a gear along its pitch circle. It is also known as *index error*. Sometimes, the *adjacent pitch error*, which is the difference between two adjacent pitches on the pitch circle, may also be evaluated in special cases. *Runout* describes the inaccuracy in radial location of the gear teeth with reference to the pitch circle. It is the maximum difference between the nominal or theoretical radial position of all teeth to the actual measured position.

Both pitch errors and runout may lead to structural integrity problems. Runout results in accumulated pitch deviation, leading to nonuniform motion, that causes transmission inaccuracy [6]. Runout has a cascading effect that may affect most other gear quality parameters including involute and/or tooth form, index and/or pitch error, lead deviation, and noise and vibration. Kinematic errors in machine tools, gear blanks, and mounting errors are the primary causes of location errors [7].

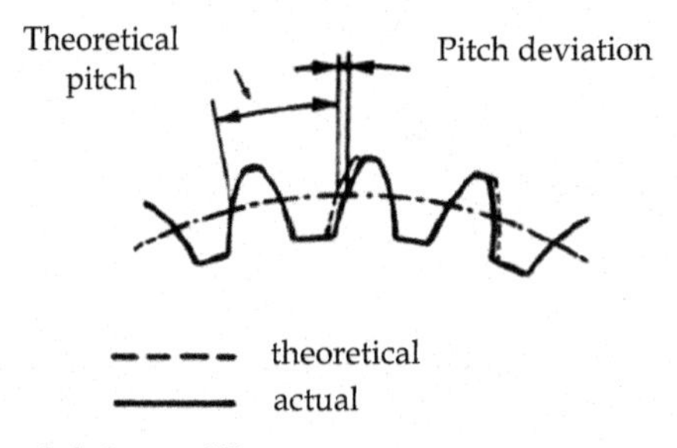

FIGURE 7.7 Concept of pitch error [5].

7.2 GEAR TOLERANCES AND STANDARDS

A tolerance may be defined as the permissible variation of a dimension or control criterion from the specified or standard value [2]. The tolerance establishes a necessary permissible range of acceptance of variation. A variation beyond the tolerance in gears leads to nonconformance and the gear is said to be "in error." It is worth mentioning that the word "deviation" is used interchangeably with variation throughout this chapter for the description of the departure from specified dimension.

In order to achieve a closer tolerance thereby ensuring high functional performance of the gears, gear fabrication should be more precise. This inevitably results in increased manufacturing cost. The fabrication needs for precision dictate the tolerances to be achieved for fulfillment of the intended functional requirement. To achieve acceptable quality with tight tolerances, the gears have to be sent for postfinishing treatments, especially when fabricated by conventional processes. In case of advanced processes, acceptable quality may be achieved by manufacturing at optimum machine tool settings.

Various international standards exists that provide different quality level/grades for all types of gears by comparing their deviations from the tolerance specified as for each quality level. The quality grade is selected based upon the functional requirement of the gear. The manufacturing technique and process is then chosen to achieve the tolerance specified for that grade.

The American Gear Manufacturers Association (AGMA) standard is an example of such an international standard. They are developed by the AGMA and approved by the American National Standards Institute (ANSI). The current AGMA standards are as follows [8,9]:

For cylindrical gears

- AGMA 915-1-A02 Inspection Practices—Part 1: Cylindrical Gears—Tangential Measurements.
- AGMA 915-2-A05 Inspection Practices—Part 2: Cylindrical Gears—Radial Measurements.
- ANSI/AGMA 2015-1-A01 Accuracy Classification System—Tangential Measurements for Cylindrical Gears.
- ANSI/AGMA 2015-2-A06 Accuracy Classification System—Radial Measurements for Cylindrical Gears.

For bevel and worm gears

- AGMA ISO 10064-6-A10 Code of Inspection Practice—Part 6: Bevel Gear Measurement Methods.
- ANSI/AGMA ISO 17485-A08 Bevel Gears—ISO System of Accuracy.
- ANSI/AGMA 2011-A98 Cylindrical Worm gearing Tolerance and Inspection Methods.

These standards cover tolerances and measuring methods. These are the current new standards that replaced the older standards ANSI/AGMA 2000-A88 (for spur and helical gears), and AGMA 390.3a (for bevel and worm gears).

To align with the other international gear standards, the accuracy grade numbers in the new standard ANSI/AGMA 2015-1-A01 were reversed (i.e., a smaller grade number represents a smaller tolerance value and as such a higher quality gear) from the previous standard ANSI/AGMA 2000-A88. The new standard ANSI/AGMA 2015-1-A01 provides 10 accuracy grades numbered A2–A11, in order of decreasing precision, whereas the old standard includes 13 quality classes numbered Q3–Q15, in order of increasing precision. Furthermore, in the new standards, the accuracy grades are segregated into three groups, i.e., high (A2–A5), medium (A6–A9), and low (A10–A11) accuracy. Corresponding to each group, the required accuracy parameters that must be met for qualification of the gear are (1) high accuracy gears—accumulated pitch, total profile and lead, form profile and lead, angle profile and lead; (2) medium accuracy gears—accumulated pitch, single pitch, and total profile and lead; and (3) low accuracy gears—accumulated pitch and single pitch.

Other important international standards are also widely used. These include the German standards DIN 3962 and 3963 for spur and helical gears [10] and DIN 3965 for bevel gears [11]; Japanese standards JIS B 1702 for spur and helical gears [12]; and JIS 1704 for bevel gears [13]; International Standards Organization ISO 1328 [14], British standards BS 436 [15].

7.3 MEASUREMENT OF GEAR ACCURACY

Essentially, there are two major classes of gear accuracy measurement:

1. Analytical inspection or measurement and
2. Functional inspection or measurement

7.3.1 Analytical Gear Inspection

Analytical gear inspection is used by gear manufacturers to determine if a gear meets the geometric specifications as described by a set of appropriate dimensions. The results of an analytical inspection may point toward the source of a problem in the gear machining process. From a diagnostic standpoint, the analytical inspection can illustrate the extent of error that would be attributed individually to the various gear tooth parameters. The parameters commonly checked by analytical gear inspection are profile error, lead error, pitch error, runout, and tooth thickness variation. This method does not detect defects (errors) such as burrs or nicks.

7.3.1.1 Macrogeometry Measurement

Three different techniques are commonly used to inspect the tooth thickness. These are measurement of the chordal thickness; over balls, pins or wires measurement; and span measurement. Typically during the process of gear manufacturing, the size of a gear is monitored throughout the production run as a means of process control. The most frequently used instrument for measuring gear tooth thickness is the gear tooth Vernier calliper which measures the chordal tooth thickness [2]. Micrometers with balls or pins/rollers are also used to measure tooth thickness. For larger gears a span measurement over several gear teeth (by base tangent method) is done with the use of plate micrometer. The actual value of the thickness is compared with its theoretical value and the difference is reported as an error.

In span measurement, a plate micrometer is used to measure the distance across several teeth along a line tangent to the base circle (i.e., referred to as the "base tangent") of the gear (see Fig. 7.8). This method overcomes the accuracy limitations of tooth thickness measurement by gear tooth Vernier calliper due to the interdependency of the double Vernier scale reading and inaccuracies introduced using the edge of the measuring jaws. The base tangent length is the distance between two parallel planes which are tangential to the opposite tooth flanks. The number of teeth over which the measurement is to be made for a particular gear is based on its total number of teeth and pressure angle and is usually available from gear manuals/handbooks. The distance measured as the base tangent length is the sum of the normal circular tooth thickness on the base circle, and the normal base pitch (number of teeth − 1). The base pitch is equal to the circumference of the base circle divided by the number of teeth. Hence, the tooth thickness can be determined based on base circle pitch and measured span.

Fig. 7.9A,B illustrates the tooth thickness measurement of an external spur gear by the over pins and balls method. In this method, two cylindrical pins/rollers, or balls of a specified diameter, are placed in diametrically opposite tooth spaces (for even numbered gear teeth), or the nearest to it (for odd numbered gear teeth), and the overall dimension W, as shown in

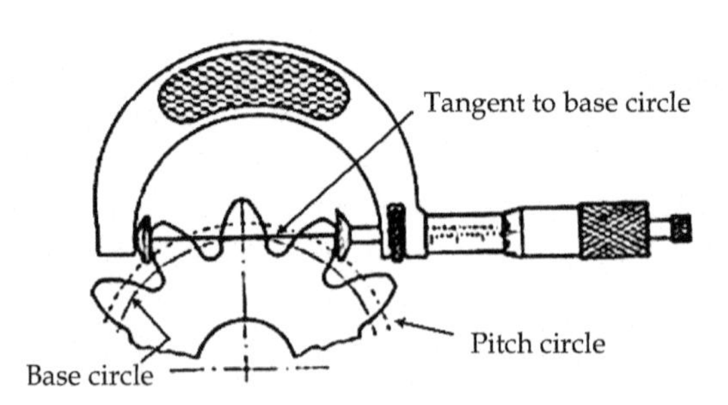

FIGURE 7.8 Span measurement using plate micrometer.

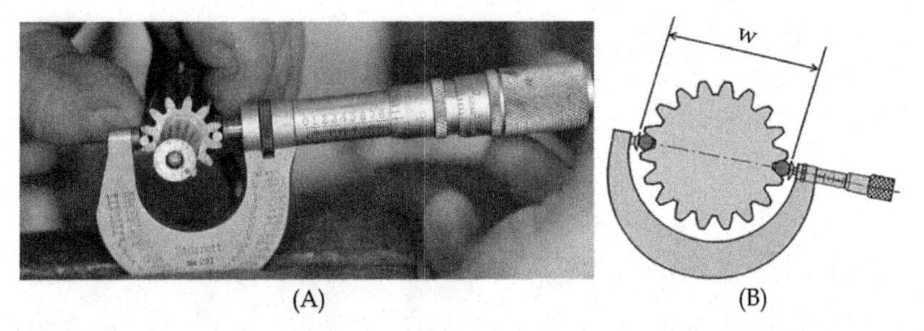

FIGURE 7.9 Tooth thickness measurement (A) by micrometer with two cylindrical pins and (B) by micrometer with two balls.

Fig. 7.9B, is measured using the micrometer. For a specified tooth thickness, the overall dimension can also be found with the help of standard formulae specified for even as well as odd number of teeth. The pin diameter is based on the diametrical pitch (or module) of the gear and whether the gear is internal or external.

7.3.1.2 Microgeometry Measurement

There are mainly two major categories of instruments used for analytical inspection of errors in microgeometry of gears. The first is traditional mechanical generative instruments, and the second is computer numerical control (CNC) instruments or coordinate measuring machines (CMM) [16]. A wide range of mechanical gear testers are available in the form of single-purpose instruments to individual inspection of profile, lead, and spacing errors; and multipurpose instrument capable of inspection of all parameters on one machine. The mechanical type hand-operated instruments which are capable of measuring/inspecting all parameters (i.e., errors in profile, lead, and pitch; and runout) typically consist of a master base circle disk, sine bars, levers, and formers that are used to generate the theoretical motion by rotating the test gear. The deviation between the actual tooth surface and the theoretical shape is then measured by a probe. The quality of the gear is then established by measuring and comparing against the tolerances as specified in the standards. The deviation reading is displayed on dial indicator, or an electrical pickup, and may be recorded on chart paper or electronically. Currently, the latest designs of inspection machines incorporate computer-controlled kinematics that ensure accuracy and flexibility in the measurement and presentation of results to reveal the topological condition of a gear in graphic, graph, or tabulated form.

CNC gear metrology machines or gear testers offer high resolution and accuracy and are extensively utilized at present to inspect gear accuracy as related to macro- and microgeometry. These are dedicated CMM that measure the geometry by dragging a special spherical-shaped probe across a

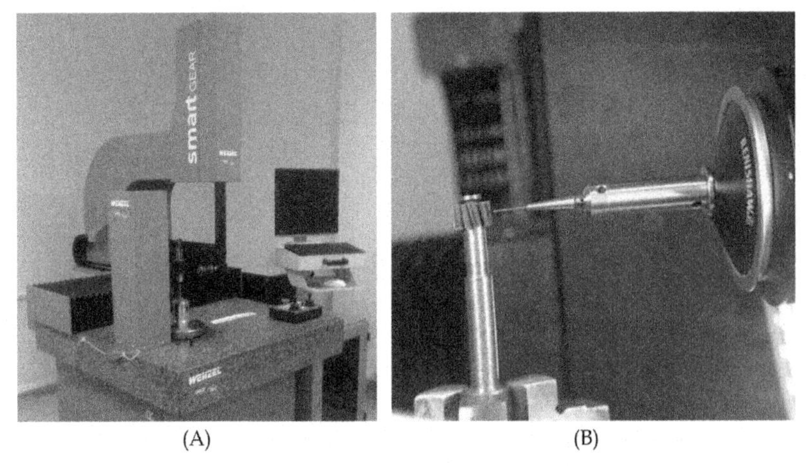

(A) (B)

FIGURE 7.10 (A) CNC gear tester (WENZEL GearTec GmbH-Germany, Gear Metrology Machine at IIT Indore-India) and (B) accuracy inspection of external helical microgear.

surface. The kinematics of the machines allow for accurate linear and angular motion. With the help of sophisticated software, it is possible to inspect cylindrical gears, worm and worm gears, conical gears, hobs, and shaving cutters all on one machine. Fig. 7.10 presents a typical example of such a system along with a typical setup to measure a microgear.

Inspection and measurement of the geometric accuracy of cylindrical gears by advanced CNC gear tester usually implies the following: The very first step is to inspect the alignment of the machine for squareness, parallelism, runout of centers and spindle, etc. It is usually required that the gear axis is aligned with one of the three main machine axes. Machine calibration using a certified master gear is also an essential step to establish accuracy and confidence levels and should be done on regular basis. Once programmed, the alignment process and other movements of the machine is stored and recalled as required for similar inspections on another different lot of the same gear. Accuracy, flexibility, and time saving are the major advantages of CNC gear testers over traditional mechanical testers.

7.3.1.2.1 Profile and Lead Measurement

In practice, the gears are mounted on a shaft or fixtures and are mounted vertically between centers or accommodated in a rotating chuck affixed to the measurement table. A single measuring probe is used for both profile and lead measurement. For profile measurement, as shown in Fig. 7.11A, the measurement probe is aligned on the test gear in the middle of the gear face. The probe is positioned against a tooth flank and traces the tooth perpendicular to the gear axis from the root to the tip or tip to the root as the part

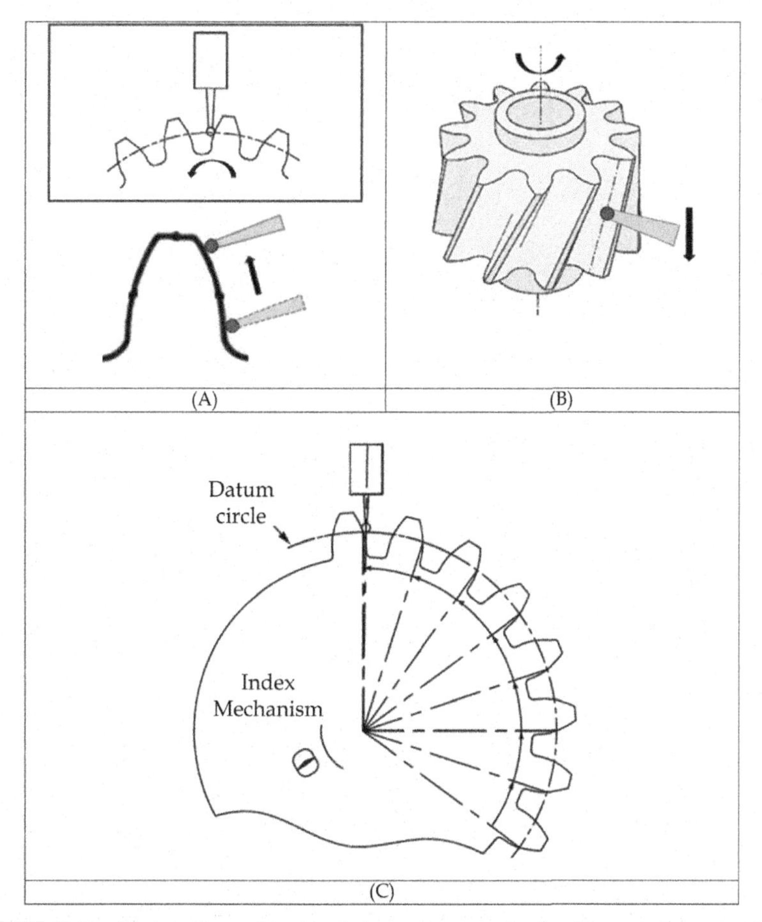

FIGURE 7.11 The position of probe during measurement of microgeometry parameters (A) profile measurement, (B) lead measurement, and (C) pitch and runout measurement.

rotates slowly. The machine compares the actual gear profile to the reference profile created against the gear specifications and may record the deviation graphically on a chart as shown in Fig. 7.12. Normally, both flanks of three to four equally spaced teeth around the gear are measured for profile and lead. The gear is assigned a quality grade as per the tolerance specified based on the reference standard.

The typical profile error chart as shown in Fig 7.12 depicts profile errors or deviations for four equally spaced teeth of an external helical gear (with specification of no. of teeth—12, pitch circle diameter—8.4 mm, and face width—5 mm). This gear also has no intentional profile modification. The profile deviations for the left flanks are located at the left side and for right

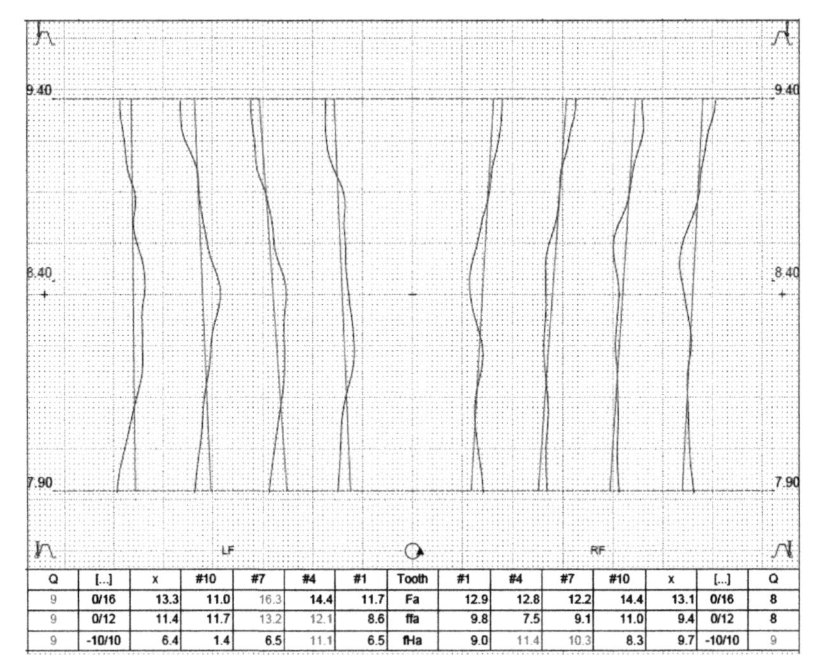

Q	[...]	x	#10	#7	#4	#1	Tooth	#1	#4	#7	#10	x	[...]	Q
9	0/16	13.3	11.0	16.3	14.4	11.7	Fa	12.9	12.8	12.2	14.4	13.1	0/16	8
9	0/12	11.4	11.7	13.2	12.1	8.6	ffa	9.8	7.5	9.1	11.0	9.4	0/12	8
9	-10/10	6.4	1.4	6.5	11.1	6.5	fHa	9.0	11.4	10.3	8.3	9.7	-10/10	9

FIGURE 7.12 Typical profile chart for an external helical gear.

flanks are located at the right side of the chart. The correct profile, i.e., theoretical perfect involute are presented as the vertical dotted lines of the chart. The irregular black traces (at the left and right) are the actual profile traces whereas the blue (light gray in print versions) traces represent a least-squares-fit of the actual data within a specified evaluation zone. The evaluation zone starts at the SAP point and ends at the EAP point. For the example gear, as shown in Fig. 7.12, the SAP and EAP points are associated with the dimensions 7.9 and 9.4, respectively. Flank profiles with excess material result in a positive deviation, while insufficient material results in a negative deviation. The numerical values for the profile errors (total profile error "F_a," profile form error "ff_a," and profile slope error "fH_a") corresponding to all the flanks are given at the bottom of the chart along with their averages (column x). The quality grades/numbers with respect to all the averaged profile deviations are also presented at the extreme left and right at the bottom of the chart.

Lead or helix measurement determines the accuracy of the teeth along their face width. It reveals the accuracy of the gear-tooth profile in the axial direction [2]. For lead measurement, the probe is positioned against a tooth flank on the pitch circle diameter point at the end of the face width (see Fig. 7.11B). The probe traces the tooth parallel to the axis and normal to the tooth surface along the face width and the deviation is recorded and graphically displayed on a chart.

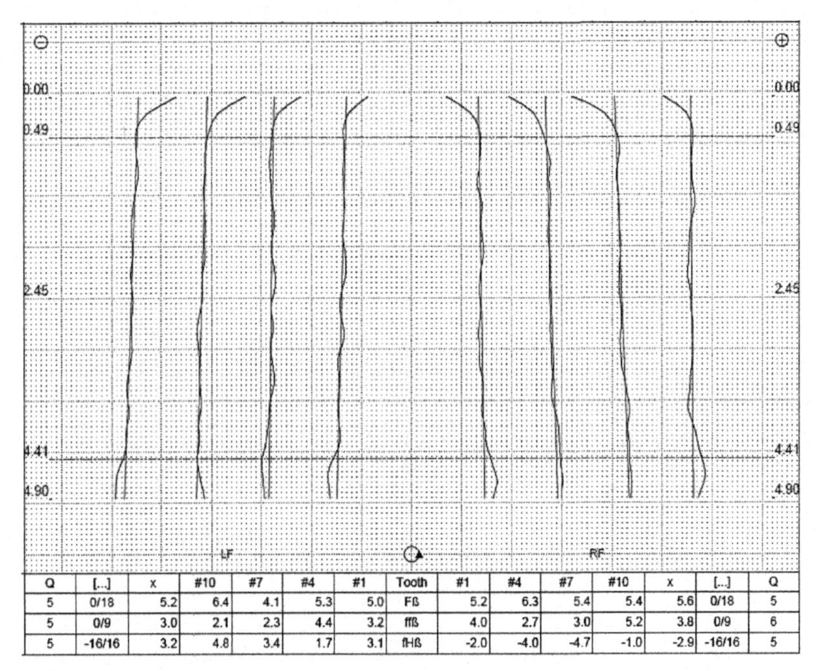

Q	[...]	x	#10	#7	#4	#1	Tooth	#1	#4	#7	#10	x	[...]	Q
5	0/18	5.2	6.4	4.1	5.3	5.0	Fß	5.2	6.3	5.4	5.4	5.6	0/18	5
5	0/9	3.0	2.1	2.3	4.4	3.2	ffß	4.0	2.7	3.0	5.2	3.8	0/9	6
5	-16/16	3.2	4.8	3.4	1.7	3.1	fHß	-2.0	-4.0	-4.7	-1.0	-2.9	-16/16	5

FIGURE 7.13 Typical lead chart for an external helical gear.

A typical lead error chart, as shown in Fig. 7.13, records from just before the end of the face width to the opposite end of the tooth along the actual probe travel direction. Similarly to the profile chart, the irregular black traces represent the actual data, whereas the blue (light gray in print versions) traces are least-squares-fitted traces of the data traces. A true helix would appear as a vertical line on the chart aligned with the dotted grid lines. The actual measured values (total lead error "F_β," lead form error "ff_β," and lead slope error "fH_β") along with their averaged values and the subsequent quality grade is displayed at the bottom of the chart.

7.3.1.2.2 Pitch and Runout Measurement

For measurement of pitch and runout, the probe is initially brought into contact with any tooth flank on the reference circle diameter point at mid face (Fig. 7.11C); this initial flank is considered as the datum tooth flank. The probe is then retracted from the tooth space, and the gear is indexed by an angle as appropriate for one tooth or pitch. The probe then moves back into the reference circle diameter of the next tooth flank, and its location is recorded. This process is repeated for a full rotation. The same procedure is conducted simultaneously for the opposite tooth flanks. Gear runout may also be inferred from these measurements.

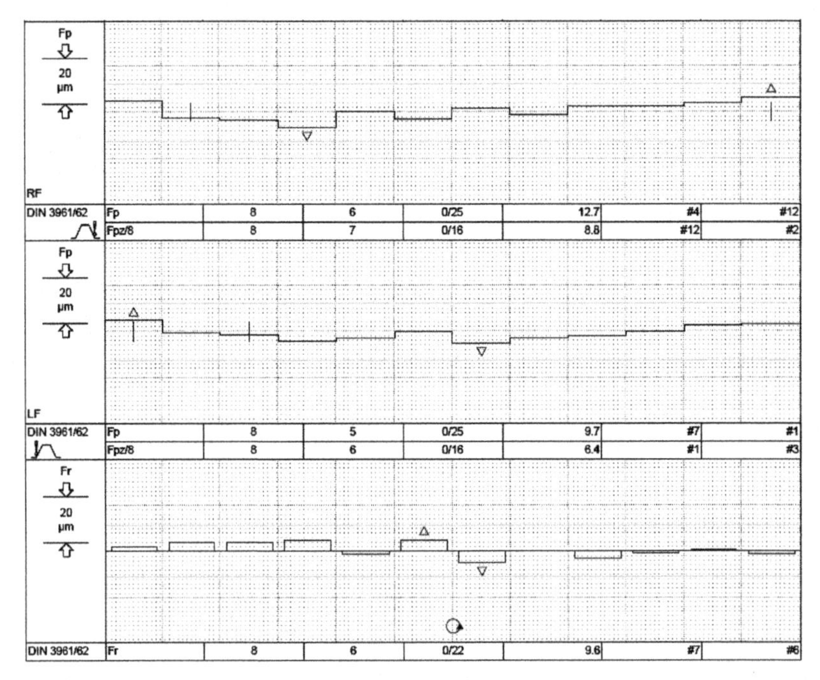

FIGURE 7.14 Typical pitch and runout chart.

The recorded results may be displayed a chart displaying single pitch error "f_p," adjacent pitch error "f_u," accumulated pitch error "F_p," and runout "F_r." A typical example of a chart of index readings that presents the accumulated pitch error for right and left flanks and runout for all the teeth is presented in Fig. 7.14.

7.3.2 Functional Gear Inspection

Functional or composite gear inspection is a qualitative method of evaluating gear accuracy where the main objective is to compare a gear to the required specifications as provided by a reference gear. Essentially, the results of functional gear inspection reveals if a gear will work as intended. This method involves rolling two gears of the same specification together (where one is a master or reference gear and other is a work gear whose quality is to be evaluated) and measuring the resultant motion to determine composite error, tooth-to-tooth error, transmission error, etc. The gears can also be tested in pairs instead of using a master gear. Functional inspection may be subdivided into two basic types:

1. double-flank inspection and
2. single-flank inspection.

The major differences between these two methods are as follows [16–19]:

- Single-flank inspection implies that only one flank is in contact during gear rolling, whereas in double-flank inspection, rolling occurs such that both flanks (right and left flanks) are in contact.
- Double-flank inspection allows for center distance variation (therefore, also referred to as the "variable center distance method"), whereas in single-flank inspection, the center distance remains fixed (therefore also known as "fixed center distance method").
- Single-flank inspection evaluates transmission errors whereas double-flank inspection cannot detect angular tooth position defects and thus cannot evaluate transmission errors.
- Double-flank gear roll testers are usually manually operated and thus relatively inexpensive.

Single-flank inspection is typically recognized as more accurate and productive (faster) than double-flank inspection and is mainly used for the inspection of high precision gears [2]. The instruments used for these tests are referred to as roll testers. It is available both in manual and computerized models. Typically, for the manual machines, rotation of the master gear occurs by hand, whereas an electric motor is used for the computerized testers. Computerized testers are typically faster and more accurate.

7.3.2.1 Double-Flank Inspection

In double-flank inspection, the gears are rolled closely meshed such that there is double-flank contact (see Fig. 7.15) and therefore no backlash. Fig. 7.16 presents a schematic of double-flank inspection along with an example of a typical double-flank roll tester.

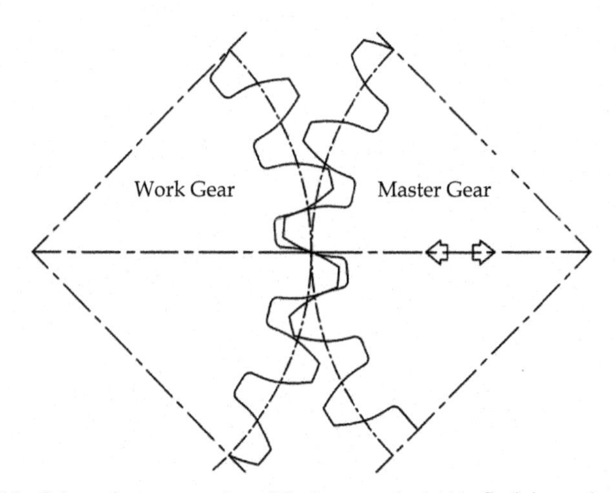

FIGURE 7.15 Schematic representation of flank contact in double-flank inspection.

(A)

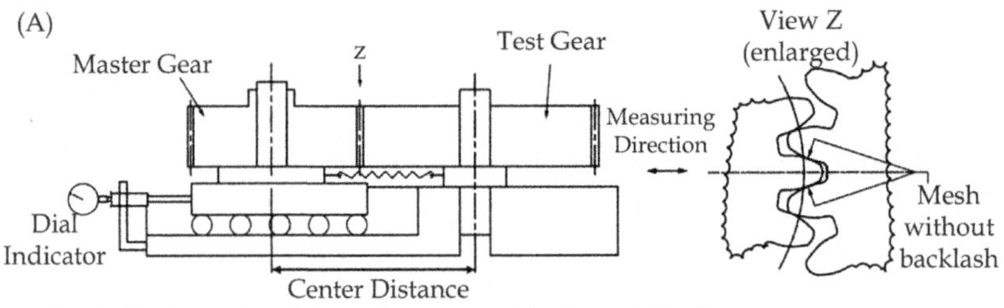

(B)

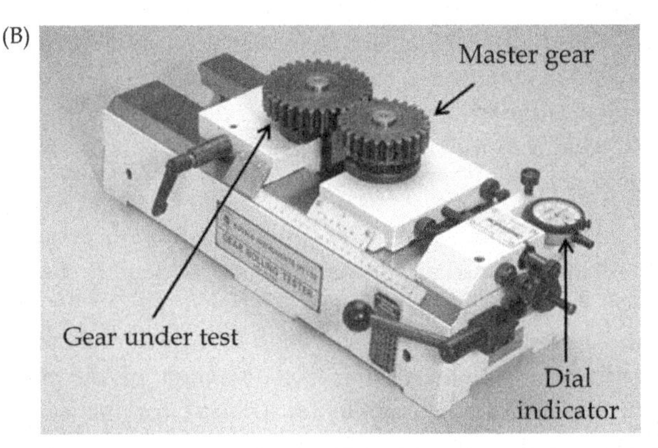

FIGURE 7.16 (A) Schematic of the double-flank inspection technique and (B) example of a typical double-flank type roll tester. *Courtesy- Kudale Instruments (P) Ltd., Pune-India.*

As shown in Fig. 7.16, the work or test gear is mounted on a fixture which allows only rotary motion of the gear, whereas the master gear is mounted on a fixture that allows linear motion along the gear centerline [2,17]. A spring with a preset force between the test gear and master gear ensures double-flank contact and zero backlash. As the master gear is rotated (either by hand or by motor), the test gear follows. During this gear rolling, the variation in the center distance occurs due to inaccuracies in the test gear that is then recorded graphically on a strip chart (in computerized testers) or displayed on a dial indicator (in manual testers).

Fig. 7.17 presents the format of a typical strip chart plotted by a double-flank tester for one complete revolution of the work gear. Double-flank inspection is used to determine runout, tooth-to-tooth rolling action, and to detect nicks and burrs in spur, helical, bevel, and worm gears [2,17]. Total composite error as shown in the chart is a consequence of full rotation, while rotation only through one pitch gives tooth-to-tooth error. The total composite error is the difference between the highest and lowest points on the graph or between the maximum and minimum readings on the dial indicator during one rotation cycle of the test gear. This includes effects of profile and helix variation, pitch errors, and runout. The tooth-to-tooth error is defined as the

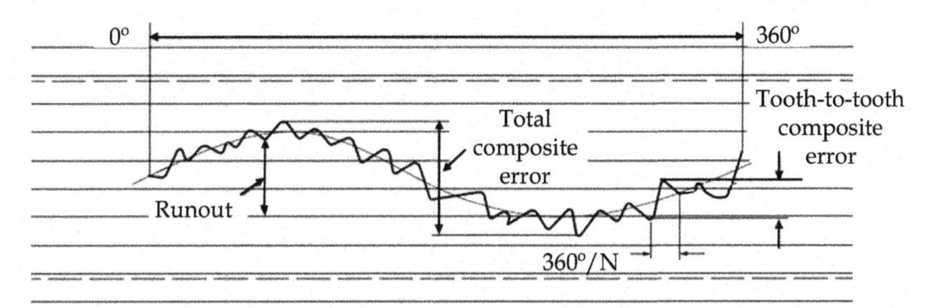

FIGURE 7.17 Graphical representation of errors in gear quality obtained by a double-flank tester.

greatest deviation within a single circular tooth pitch. It shows the worst tooth on the entire gear. The test gear is rejected if the total composite variation is greater than the allowable total composite tolerance as defined during gear design. Double-flank inspection does not provide information regarding accumulated pitch variation or specific tooth profile characteristics.

7.3.2.2 Single-Flank Inspection

The single-flank inspection method can evaluate all elements of gear quality, such as profile conjugacy (closely related to typical gear noise), adjacent pitch variation, accumulated pitch variation, and runout except lead (tooth alignment) or helix deviation. Moreover, the transmission error may be directly measured. Transmission error is the main cause of gear noise and is defined as the deviation of the driven gear, for a given angular position of the driving gear, from the position that the driven gear would occupy if the gears were geometrically perfect. For applications that require positional accuracy, it is important to evaluate the transmission error [2]. This method is also used to evaluate the effect of combined profile errors on bevel gears with greater confidence than by using tooth-contact pattern testing [18].

Single-flank inspection implies that the mating gears roll together on a fixed center distance with backlash and in such a way that only one pair of flanks makes contact (as shown in Fig. 7.18). Compared to other methods, single-flank inspection more closely resembles the actual operation of rotating gears while appropriate quality parameters can be measured.

Fig. 7.19 presents the basic operational principle of a single-flank inspection machine (tester).

Minute changes in angular velocity and angular position of the individual meshed gear pair are measured and compared by circular optical gratings or encoders. These encoders are fixed to each gear shaft as shown in Fig 7.19. Each encoder provides a stream of pulses. The frequency of these pulses relate to the angular velocity. Differences in angular velocity between the two gears are picked up by a phase comparator and relates to errors in the test gear. This difference is recorded as an analog waveform on a strip chart

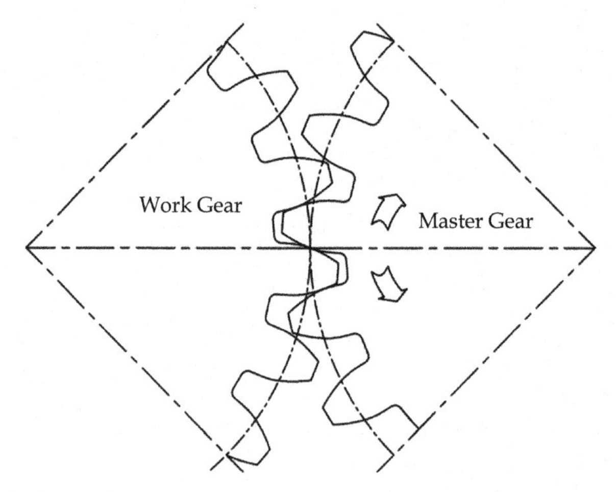

FIGURE 7.18 Schematic representation of flank contact in single-flank inspection.

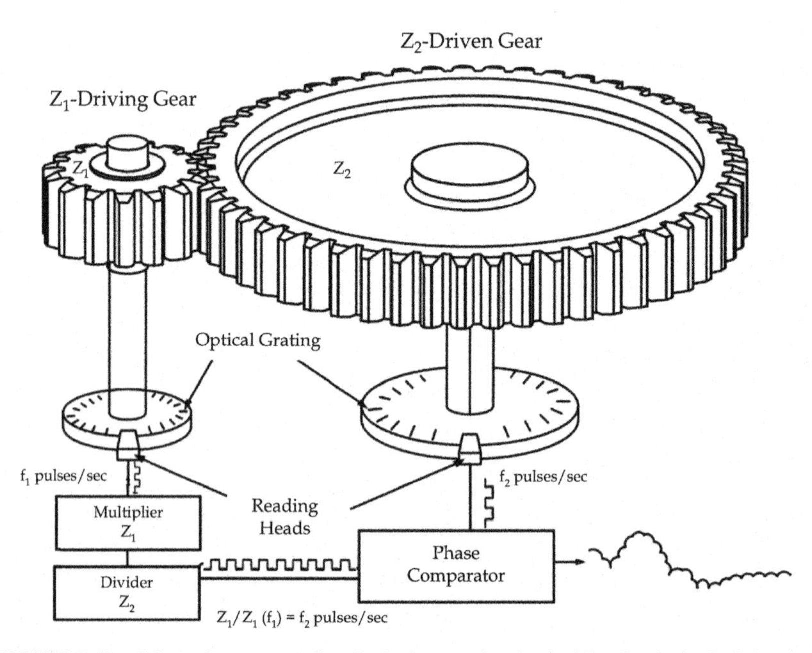

FIGURE 7.19 Schematic representation the basic operational principle of a single-flank inspection machine.

generated by the instrument (see Fig. 7.20). Phase differences of less than one arc-second can easily be detected by this method.

The waveform or curve displayed in this chart is for one revolution of a test gear. This curve is known as a "total transmission error curve" and composed

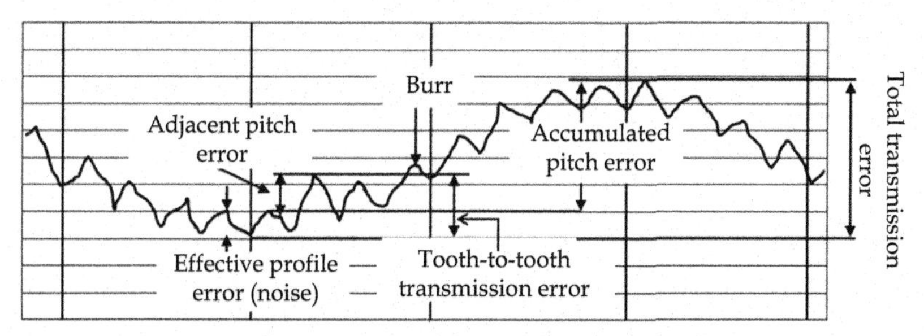

FIGURE 7.20 Typical output of a single-flank tester for evaluation of gear quality.

of various quality parameters such as effective profile error, burr amplitude, adjacent pitch error, accumulated pitch error, tooth-to-tooth composite error, and total composite error [19]. The total composite error is the difference between the highest and lowest points on the graph within one revolution of the test gear and includes the effect of a portion of profile error and accumulated pitch error. The tooth-to-tooth error is the variation in transmission error at the tooth meshing frequencies and mainly the consequence of profile errors and single pitch errors. These are important for the evaluation of gear noise. Single-flank inspection is usually conducted at relatively low loads.

This chapter has presented various geometry parameters pertaining to gear accuracy, and aspects of their analytical and functional evaluation. It introduced advanced techniques such as accuracy evaluation by CNC gear tester, and single-flank and double-flank inspection by roll testers. Continuous development is ongoing as regards to gear measurement with increased accuracy, flexibility, and application to a wider range of different types of gears. Measurement techniques related to gears as regards to surface finish, residual stress, hardness, and microstructure are also ongoing for gears from micro to giant sized. Various portable, custom-made, and on-site inspection machines/instruments are available to provide solutions to the shop floor or laboratory based inspection requirements.

REFERENCES

[1] K. Gupta, N.K. Jain, Chapter 1: Introduction, in Near-Net shape manufacturing of miniature spur gears by wire spark erosion machining, Springer Science and Business Media Pvt Ltd., Singapore, 2016, pp. 1–15.

[2] D.P. Townsend, Dudley's Gear Handbook, Tata McGraw-Hill Publishing Company, New Delhi, 2011.

[3] K. Gupta, N.K. Jain, Deviations in geometry of miniature gears fabricated by wire electrical discharge machining, in: Proceedings of International Mechanical Engineering Congress & Exposition (IMECE 2013) of ASME, V010T11A047, 13–21 November 2013, San Diego, USA, 2013.

[4] E. Lalwson, The basics of gear metrology and terminology, part I, Gear Technology Magazine, September/October 1998, pp. 41–50.

[5] K. Gupta, N.K. Jain, Analysis and optimization of micro-geometry of miniature gears manufactured by wire electric discharge machining, Precis. Eng. 38 (4) (2014) 728–737.

[6] R.E. Smith, What is runout. And by should I worry about it, Gear Technology Magazine, January/February 1991, pp. 43–44.

[7] E. Lawson, The basics of gear metrology and terminology, part II, Gear Technology Magazine, November/December 1991, pp. 67–74.

[8] American Gear Manufacturers Association, Technical Publications Catalog, Technical Resources, April 2014.

[9] D. Gimpert, A new standard in gear inspection, Gear Solutions Magazine, October 2005, pp. 34–38.

[10] DIN 3962 and 3963 Standards, Tolerances for Cylindrical Gear Teeth, © Beuth Verlag GmbH, Deutsche Normen, Berlin, Germany, 1978.

[11] DIN 3965 Standard, Tolerances for bevel gears, © Beuth Verlag GmbH, Deutsche Normen, Berlin, Germany, 1979.

[12] JIS B1702 Standard. Accuracy for spur and helical gears, Japanese Industrial Standard, 1976.

[13] JIS B1704 Standard. Accuracy for bevel gears, Japanese Industrial Standard, 1978.

[14] ISO 1328 Standard. System of accuracy for parallel involute gears, International Organization for Standardization, 1975.

[15] BS 436 Standard. Specification for spur and helical gears, British Standards Institute, 1967.

[16] R.E. Smith, Quality gear inspection, part II, November/December 1994, Gear Technology Magazine, pp. 31–34.

[17] D. Gimpert, An elementary guide to gear inspection, Gear Solutions Magazine, June 2005, pp. 32–38.

[18] R.E. Smith, Single-flank testing of gears, Gear Technology Magazine, May/June 2004, pp. 18–21.

[19] R.E. Smith, Identification of gear noise with single flank composite measurement, Gear Technology Magazine, May/June 1986, pp. 17–29.

Index

Note: Page numbers followed by "*f*" and "*t*" refer to figures and tables, respectively.

A

B

C

D

H

I

N

O

P

T

U